Automatic Differentiation in MATLAB Using ADMAT with Applications

SOFTWARE • ENVIRONMENTS • TOOLS

The SIAM series on Software, Environments, and Tools focuses on the practical implementation of computational methods and the high performance aspects of scientific computation by emphasizing in-demand software, computing environments, and tools for computing. Software technology development issues such as current status, applications and algorithms, mathematical software, software tools, languages and compilers, computing environments, and visualization are presented.

Software, Environments, and Tools

Thomas F. Coleman and Wei Xu, *Automatic Differentiation in MATLAB Using ADMAT with Applications*

Walter Gautschi, *Orthogonal Polynomials in MATLAB: Exercises and Solutions*

Daniel J. Bates, Jonathan D. Hauenstein, Andrew J. Sommese, and Charles W. Wampler, *Numerically Solving Polynomial Systems with Bertini*

Uwe Naumann, *The Art of Differentiating Computer Programs: An Introduction to Algorithmic Differentiation*

C. T. Kelley, *Implicit Filtering*

Jeremy Kepner and John Gilbert, editors, *Graph Algorithms in the Language of Linear Algebra*

Jeremy Kepner, *Parallel MATLAB for Multicore and Multinode Computers*

Michael A. Heroux, Padma Raghavan, and Horst D. Simon, editors, *Parallel Processing for Scientific Computing*

Gérard Meurant, *The Lanczos and Conjugate Gradient Algorithms: From Theory to Finite Precision Computations*

Bo Einarsson, editor, *Accuracy and Reliability in Scientific Computing*

Michael W. Berry and Murray Browne, *Understanding Search Engines: Mathematical Modeling and Text Retrieval, Second Edition*

Craig C. Douglas, Gundolf Haase, and Ulrich Langer, *A Tutorial on Elliptic PDE Solvers and Their Parallelization*

Louis Komzsik, *The Lanczos Method: Evolution and Application*

Bard Ermentrout, *Simulating, Analyzing, and Animating Dynamical Systems: A Guide to XPPAUT for Researchers and Students*

V. A. Barker, L. S. Blackford, J. Dongarra, J. Du Croz, S. Hammarling, M. Marinova, J. Wasniewski, and P. Yalamov, *LAPACK95 Users' Guide*

Stefan Goedecker and Adolfy Hoisie, *Performance Optimization of Numerically Intensive Codes*

Zhaojun Bai, James Demmel, Jack Dongarra, Axel Ruhe, and Henk van der Vorst, *Templates for the Solution of Algebraic Eigenvalue Problems: A Practical Guide*

Lloyd N. Trefethen, *Spectral Methods in MATLAB*

E. Anderson, Z. Bai, C. Bischof, S. Blackford, J. Demmel, J. Dongarra, J. Du Croz, A. Greenbaum, S. Hammarling, A. McKenney, and D. Sorensen, *LAPACK Users' Guide, Third Edition*

Michael W. Berry and Murray Browne, *Understanding Search Engines: Mathematical Modeling and Text Retrieval*

Jack J. Dongarra, Iain S. Duff, Danny C. Sorensen, and Henk A. van der Vorst, *Numerical Linear Algebra for High-Performance Computers*

R. B. Lehoucq, D. C. Sorensen, and C. Yang, *ARPACK Users' Guide: Solution of Large-Scale Eigenvalue Problems with Implicitly Restarted Arnoldi Methods*

Randolph E. Bank, *PLTMG: A Software Package for Solving Elliptic Partial Differential Equations, Users' Guide 8.0*

L. S. Blackford, J. Choi, A. Cleary, E. D'Azevedo, J. Demmel, I. Dhillon, J. Dongarra, S. Hammarling, G. Henry, A. Petitet, K. Stanley, D. Walker, and R. C. Whaley, *ScaLAPACK Users' Guide*

Greg Astfalk, editor, *Applications on Advanced Architecture Computers*

Roger W. Hockney, *The Science of Computer Benchmarking*

Françoise Chaitin-Chatelin and Valérie Frayssé, *Lectures on Finite Precision Computations*

Automatic Differentiation in MATLAB Using ADMAT with Applications

Thomas F. Coleman
University of Waterloo
Waterloo, Ontario, Canada

Wei Xu
Tongji University
Shanghai, P. R. China
and
Global Risk Institute
Toronto, Ontario, Canada

siam.
Society for Industrial and Applied Mathematics
Philadelphia

10 9 8 7 6 5 4 3 2 1

Publisher	David Marshall
Acquisitions Editor	Elizabeth Greenspan
Developmental Editor	Gina Rinelli Harris
Managing Editor	Kelly Thomas
Production Editor	Lisa Briggeman
Copy Editor	Bruce Owens
Production Manager	Donna Witzleben
Production Coordinator	Cally Shrader
Compositor	Techsetters, Inc.
Graphic Designer	Lois Sellers

Library of Congress Cataloging-in-Publication Data

Names: Coleman, Thomas F. (Thomas Frederick), 1950- author. | Xu, Wei (Mathematician), author.

Title: Automatic differentiation in MATLAB using ADMAT with applications / Thomas Coleman, University of Waterloo, Waterloo, Ontario, Canada, Wei Xu, Tongji University, Shanghai, P.R. China, and, Global Risk Institute, Toronto, Canada.

Description: Philadelphia : Society for Industrial and Applied Mathematics, [2016] | Series: Software, environments, and tools | Includes bibliographical references and index.

Identifiers: LCCN 2016009172 (print) | LCCN 2016016788 (ebook) | ISBN 9781611974355 | ISBN 9781611974362

Subjects: LCSH: Numerical differentiation. | Mathematical optimization. | Automatic differentiations. | Financial engineering–Computer programs. | MATLAB. | ADMAT.

Classification: LCC QA355 .C59 2016 (print) | LCC QA355 (ebook) | DDC 518/.53028553–dc23

LC record available at *https://lccn.loc.gov/2016009172*

Contents

Preface

A fundamental need that occurs across the field of scientific computing is the calculation of partial derivatives. For example, to determine the direction of sharpest ascent of a differentiable function of n variables is to determine the gradient, i.e., the vector of first partial derivatives. The sensitivity of this vector, in turn, is the matrix of second partial derivatives (i.e., Hessian matrix), assuming the original function is twice continuously differentiable. Optimization of multidimensional functions, nonlinear regression, and multidimensional zero-finding represent broad and significant numerical tasks that rely heavily on the repeated calculation of derivatives. The vast majority of modern methods for solving these numerical problems demand the repeated calculation of partial derivatives. Hence, across numerous application areas, determining partial derivatives accurately and efficiently (in space and time) is of paramount importance.

Computational finance is one such area relying on derivative calculations. Determining the "Greeks" corresponding to (models of) financial instruments comes down to the determination of first derivatives. While this is an easy exercise for simple instruments, it can be a complex and expensive task for "exotic" instruments. Derivatives are required to measure sensitivity to parameters (such as initial stock price, volatility, interest rates) and subsequently determine hedging portfolios. Second derivatives are sometimes used to further augment this hedging strategy. For more discussion on the use of automatic differentiation in finance, see Chapter 7.

Engineering applications involve the differentiation of complex codes. Automatic differentiation (AD) can be applied, straightforwardly, to get all necessary partial derivatives (usually first and possibly second derivatives) regardless of the complexity of the code. However, the space and time efficiency of AD can be dramatically improved, sometimes transforming a problem from intractable to highly feasible, if inherent problem structure is used to apply AD in a judicious manner. This is an important matter, and we devote significant parts of Chapters 2, 5, 4, 7, and 8 to these issues.

The ideas and concepts presented in this book are largely independent of computer languages. AD tools now exist in many computer languages. However, we develop all our examples in the popular MATLAB environment and illustrate with the general differentiator ADMAT[1] for use in MATLAB. A version of ADMAT is available for this book from www.siam.org/books/se27. The reader is encouraged to try test examples with ADMAT. Occasionally some MATLAB codes also contain snippets of C or Fortran for a variety of reasons. ADMAT can handle such codes—as illustrated with some examples in Chapter 6—though the C or Fortran snippets are themselves approximately differenti-

[1]The user manual for ADMAT 2.0 is given in [26]. Code is available in [95].

ated via finite differencing. Generally, we recommend that the codes be 100% MATLAB if possible.

Why AD? It is our belief that AD is generally the best available technology for obtaining derivatives of codes used in scientific computing. Let us consider the other possibilities:

1. Finite differencing is by far the most common method in use today. It is popular primarily because it is easy to understand and easy to implement, requiring no additional tools. However, it is a seriously flawed approach, and we advise against its use. First, it requires a differencing parameter that is difficult to choose well in any general sense. The accuracy of the derivative depends on this choice, and a good choice can be challenging [53]. AD, in contrast, requires no such parameter and is highly accurate. In addition, finite differencing can require considerably more computing time compared to AD. An example of this can be seen in gradient computation (see Chapter 1 for more details).

2. Hand coding of the derivative functions is also a popular approach. If done correctly, this option can be useful yet laborious. Moreover, it is incredibly easy to err in hand coding derivative functions and notoriously hard to find the error. In addition, hand-coded functions cannot match the speed of AD in many cases (see Chapter 1 for examples), whereas in principle AD cannot be outperformed by hand-coded derivative functions.

3. Symbolic differentiation is a technology that differentiates a program at symbol-by-symbol, independent of iterate values, and produces code for the derivatives [76]. However, this approach has two serious drawbacks. First, it can be very costly and therefore appears not to be competitive for "large-scale" problems. Second, and more subtly, the derivative code that is produced may actually represent a poor method for evaluating the derivative in terms of round-off error magnification.

In summary, AD has strong advantages over alternative ways to compute derivatives. Many of the "too much space" and "too much time" criticisms of several years ago have been overcome with recent technical advances.

Automatic Differentiation Background

AD has a long history: Newton and Leibniz applied derivative computation in differential equations instead of symbolic formulae. In 1950, compilers were developed to compute derivatives by the chain rule. There are several early independent contributors to the field of automatic differentiation. Iri et al. [66, 67, 68], Speelpenning [92], and Rall [84, 85, 86, 87, 88, 89, 90] are three of the AD pioneers. In our view Andreas Griewank, Chris Bischof, and their colleagues can be credited for much of the recent interest in AD as well as the reemergence of AD as a very practical tool in scientific computing; see e.g., [4, 55, 57].

There are two basic modes of AD: the forward mode and the reverse mode. The forward mode is a straightforward implementation of the chain rule. The earliest compilers [6, 99, 100] were implemented in this mode. Later, the potential of the reverse mode was realized by several independent researchers [20, 75, 91, 92, 101]. Compared to the forward mode, the reverse mode records each intermediate operation of the differentiating function on a "tape," then rolls back from the end of the "tape" to the beginning to obtain the derivative. When computing a gradient for a large problem, reverse mode requires much lower complexity than the forward mode, but it also requires significantly more memory. Reverse mode versus forward mode often represents a trade-off between complexity and space. In order to deal with the massive space requirement problem sometimes posed by reverse mode, considerable work has been done, roughly described under the banner of a computer science technique known as "checkpointing" [59]. Checkpointing is a general procedure that can be defined on a computational graph of a function to evaluate and differentiate an arbitrary differentiable function $z = f(x)$, where for convenience we restrict our discussion to a scalar-valued function of n-variables, i.e., $f : \mathbb{R}^n \to \mathbb{R}$. The general idea is to cut the computational graph at a finite set of places (checkpoints) in a forward evaluation of the function $f(x)$, saving relevant state information at each of those checkpoints, and then recompute the information between checkpoints in a backward sweep in which the derivatives are computed. The advantage is potentially considerable savings in required space (compared to a reverse-mode evaluation of the gradient, which requires saving the entire computational graph). There is some computing cost, but the total cost is just a constant factor times the cost to evaluate the function (i.e., the cost to perform the forward sweep). So this checkpointing procedure is the same order of work as straight reverse-mode AD but uses considerably less space.

There are many ways to proceed with this general checkpointing concept—some involve machine/environment specific aspects, some involve user interference, and all involve making choices about where to put the checkpoints and what state information to store [80]. Examples of checkpointing manifestations are included in [65]. Typically the user of the checkpointing idea either has to explicitly, in a "hands-on" way, decide on these questions for a particular application or allow a checkpoint/AD tool to decide. There is no "well-accepted" automatic way (that we know of) to choose optimal checkpoints, and choice of checkpoint location clearly matters. The authors in [43] have done some work on this matter, with reference to determining the general Jacobian matrix: one conclusion is that it is very hard to choose an ideal set of checkpoints in an automatic way, given the code to evaluate $f(x)$. Some AD software apparently does decide automatically on checkpoint location (and which state variable to save), though the actual rules that are used are, in our view, somewhat opaque.

There is also considerable interest in the AD community in explicitly reformulating "macro-steps" (such as implicit equation solution, or matrix equations) as an alternative to "checkpointing," as "elemental" operations and reusing by-products of the forward step to expedite the corresponding adjoint step. An example of such a "macro-step" is a single time step in a discrete time evolution, such as an optimal control problem. One of the themes in this book is the importance of applying AD in a structured way (for efficiency), which can be regarded as a natural generalization of such an approach. Pantoja's construction of the exact Newton step leads to recurrence equations [83] very similar to those introduced in Chapter 3. Another development by Gower and Mello [54] is to use the technique of edge-pushing, which involves the exploitation of internal, implicit structural sparsity to construct the complete derivative matrix. Edge-pushing also exploits the sparsity of the working set [98] to reinforce the connection with checkpointing.

Moreover, in many applications, the derivative matrix is sparse or structured. An efficient approach for computing the derivative matrix through AD is to explore the sparsity and combine AD with a graph coloring technique. Coleman et al. [31, 32] proposed a bi-coloring method to compute a sparse derivative matrix efficiently, based on the forward and reverse modes in AD. Recently, some acyclic and star coloring methods were proposed [49, 50] to efficiently compute the second-order derivative matrix with AD. A good survey of the derivative matrix coloring methods can be found in [51]. Various sparsity pattern determination techniques and a package for derivative matrices were also developed in the AD framework [64, 97, 98]. Readers are encouraged to visit the website for the AD community at www.siam.org/books/se27. ADMAT is a toolbox designed to help readers implement the AD concepts and compute first and second derivatives and related structures efficiently, accurately, and automatically with the templates introduced in this book. This toolbox employs many sophisticated techniques, exploiting sparsity and structure, to help readers gain efficiency in the calculation of derivative structures (e.g., gradients, Jacobians, and Hessians).

Synopsis of This Book

The focus of this book is on how to use AD to efficiently solve real problems (especially multidimensional zero-finding and optimization) in the MATLAB environment. A matrix-friendly point of view is developed, and with this view we illustrate how to exploit structure and reveal hidden sparsity to gain efficiency in the application of AD. MATLAB templates are provided to help the user take advantage of structure to increase AD efficiency. The focus of earlier AD reference books, e.g., [60, 79], is somewhat different with more attention on computer science issues and viewpoints such as checkpointing, sparsity exploitation, tensor computing, operator overloading, and compiler development.

This book is concerned with the determination of the first and second derivatives in the context of solving scientific computing problems in MATLAB. We emphasize optimization and solutions to nonlinear systems. All examples are illustrated with the use of ADMAT [26].

The remainder of this book is organized as follows. Chapter 1 introduces some basic notions underpinning automatic differentiation methodologies and illustrates straightforward examples with the use of ADMAT. The emphasis in this chapter is on simplicity and ease of use: efficiency matters will be deferred to subsequent chapters. In Chapter 2 we illustrate that Jacobian-matrix, and Hessian-matrix products can be determined directly by AD without first determining the Jacobian (Hessian) matrix. Not only is this useful and cost effective in its own right, but this small generalization to basic AD also opens the window to the efficient determination of sparse derivative matrices (Jacobians and Hessians). Sparse techniques are subsequently discussed in Chapter 2.

While many problems in scientific computing are sparse, i.e., exhibiting sparse Jacobian and/or Hessian matrices, there are equally many problems that are dense but with underlying structure. Chapter 4 develops this notion of structure and illustrates a surprising but powerful result: efficient sparse techniques for the AD determination of Jacobians (Hessians) can be effectively applied to dense but structured problems. This is a far-reaching observation since many applications exhibit structure, and the ability to use sparse AD techniques on dense problems yields significant speed benefits. Chapter 5 also discusses this structure theme. It begins with the observation that the important multidimensional Newton step can be computed with AD technology without first (entirely) computing the associated Jacobian (or Hessian) matrix. This is related to this notion of structure and, again, can yield significant computational benefits. Chapter 5 develops this idea in some generality while addressing a basic question in scientific computing: how best to compute the Newton step.

In Chapter 3 we illustrate how to use ADMAT with the MATLAB optimization toolbox, with several examples. Occasionally, MATLAB users include FORTRAN and C codes with the calling MATLAB program: in Chapter 6 we show how ADMAT can "work around" such complications. Chapter 7 deals with structure problems in more detail. Specifically, inverse problems are considered with several examples, including a computational finance problem. Chapter 8 provides a template for using ADMAT on structured problems.

Appendix A discusses installation of ADMAT. In Appendix B, we cover some of the basic mechanisms behind AD.

Acknowledgments

First, we thank Arun Verma for all his preliminary work and help in developing ADMAT 1.0. In addition, we very much appreciate his advice on our development of ADMAT 2.0.

We also thank our colleagues at the University of Waterloo, Cornell University, and Tongji University for their help and advice. We are particularly grateful to our many graduate students who worked with early versions. Six of our former students specifically helped with this book and tested all the enclosed examples. They are Xi Chen, Yuehuan Chen, Wanqi Li, Yichen Zhang, and Hanxing Zhang (all at the University of Waterloo) and Ling Lu at Tongji University. Special thanks to Wanqi Li for his significant help with the design and testing of the template for structured problems described in Chapter 9.

The research behind this book has been supported over the years by the U.S. Department of Energy and the Canadian Natural Sciences and Engineering Research Council. The first author is particularly grateful to Mike and Ophelia Lazaridis for their research support.

Chapter 1
Fundamentals of Automatic Differentiation and the Use of ADMAT

The basic idea behind automatic differentiation is simple: it is the chain rule.

A vector- or scalar-valued function of several real variables, expressed as computer code, is ultimately broken down into a partially ordered sequence of simple unitary and binary mathematical operations when evaluating the function at an iterate. So, executable code, derived from the user source code to evaluate a function of several variables, is an ordered sequence of simple unitary or binary mathematical operations. There is a finite well-understood set of such unitary or binary operations (e.g., $+$, $-$, $*$, $/$, sin, cos, $\ldots$), and the derivatives of these functions, with respect to their inputs, are known and easy to compute. Therefore, it is possible to evaluate the derivative of the entire chain of operations, with respect to the original variables, by evaluating the derivative functions of each of the simple mathematical operations in turn, while evaluating the function itself. By effectively applying the chain rule, the Jacobian matrix (including the special case of the gradient) is obtained. When this is done—evaluating the derivatives in turn while the function itself is being evaluated and applying the chain rule—the resulting Jacobian is obtained by the "forward mode" of AD.

A variation on the above procedure is to evaluate the function and store the intermediate variables as they are created, thus creating the computational "tape." The tape is then processed in reverse order, again applying the chain rule, to finally obtain the Jacobian matrix with respect to the original variables. This is known as the "reverse mode" of AD. The advantage of using reverse mode may not be immediately transparent—but there are circumstances, illustrated subsequently in this book, in which it is clearly and significantly superior to the forward mode—but there also is a potential disadvantage, and that is space. In reverse mode the entire computational tape (i.e., all intermediate variables generated in the forward pass to evaluate the function) must be stored. In Chapters 5, 4, and 7, we discuss ways in which this space disadvantage can be mitigated.

In Appendix B we illustrate these two processes, forward and reverse mode, and the computational tape (as well as the associated computational graph) on a very simple example. Below we present a useful matrix view of AD.

1.1 ▪ A Matrix View of AD

There are three distinct ways of describing AD: from the computational tape perspective, through the computational graph view, and by using matrices. All three perspectives are useful. The computational tape description and the computational graph view are illustrated in Appendix B. Here we develop the matrix description.

Suppose we have a program that computes the function $z = F(x)$, $F : \mathcal{R}^n \to \mathcal{R}^m$, F is differentiable, and n, m are positive integers. The evaluation of $F(x)$ will produce an ordered sequence of intermediate variables, $y = (y_1, \ldots, y_p)$. Typically $p \gg m, n$, and each y_i is a result of a simple, differentiable, unitary, or binary operation on one or two previously determined intermediate (or original) variables. We denote the set of these simple one- or two-variable functions as *atomic* functions. Hence, we claim that $F(x)$ can be represented as a partially ordered sequence of atomic functions. That is, for each index i, y_i is an atomic function of one or two elements in the set $(x_1, \ldots, x_n, y_1, \ldots, y_{i-1})$. In other words, for any function $F(x)$, it can be decomposed into the following intermediate variables and functions:

solve y_1 :	$F_1^E(x, y_1) = 0$
solve y_2 :	$F_2^E(x, y_1, y_2) = 0$
$\vdots$	
solve y_p :	$F_p^E(x, y_1, y_2, \ldots, y_p) = 0$
solve for output z :	$z - F_{p+1}^E(x, y_1, y_2, \ldots, y_p) = 0$

In principle we can differentiate this process with respect to the independent (input) variables x and the intermediate variables y to obtain a giant $(p+m) \times (n+p)$ matrix C of partial derivatives. We note that C is very sparse since each row represents the derivatives of a unitary or binary elementary function—hence *there are at most three nonzeros per row*. If we order the variables with x preceding y, and the elements of y are ordered consistent with the partial ordering induced by the evaluation of F, then matrix C can be written as

$$C = \begin{pmatrix} A & L \\ B & M \end{pmatrix}, \tag{1.1}$$

where $A \in \mathcal{R}^{p \times n}$, $B \in \mathcal{R}^{m \times n}$, $L \in \mathcal{R}^{p \times p}$, $M \in \mathcal{R}^{m \times p}$, and L is lower triangular and nonsingular. Matrix C is the (giant) Jacobian of the nonlinear process F with respect to the input and intermediate variables x and y. The Jacobian of F with respect to x can be extracted from (1.1) via a Schur-complement computation:

$$J = B - ML^{-1}A. \tag{1.2}$$

There are two basic ways to compute (1.2). The ways differ in the order of operations. One method is

$$J = B - M[L^{-1}A]. \tag{1.3}$$

Note that (1.3) implies that the elements of matrix L can be accessed from top to bottom, which is to say the order in which they are created, since the matrix C is created top to bottom (as the function F is evaluated). Computation (1.3) corresponds to the forward mode of AD—because L and A can be accessed from top to bottom (i.e., as they are created), it follows that the computational tape is read only in the forward direction, and tape need not be stored in its entirety. It can be argued from (1.3) that the amount of work to determine J is proportional to $n \cdot p$. Of course, the work to evaluate F itself is proportional to p.

In terms of computational time and space required, the forward mode is equivalent to the well-known (and heavily used) method of finite differencing. If we denote the time to evaluate the function F at an arbitrary argument x as $\omega(F)$, then both finite differencing and forward-mode AD can compute the Jacobian in time proportional to $n \cdot \omega(F)$.

Example 1.1. Define $F(x) = \binom{3x_1+2x_2}{5x_1^2+4x_1+x_2}$ and suppose our current iterate is $\bar{x} = \binom{1}{2}$. Then it is easy to see that $F(\bar{x}) = \binom{7}{11}$, $J(\bar{x}) = \begin{pmatrix} 3 & 2 \\ 14 & 1 \end{pmatrix}$. A suitable sequence of intermediate variables y is

$$y_1 = 3x_1,\ y_2 = 2x_2,\ y_3 = x_1^2,\ y_4 = 5y_3,\ y_5 = 4x_1,\ y_6 = y_5 + x_2, \tag{1.4}$$

and finally the output is obtained:

$$z_1 = y_1 + y_2,\ z_2 = y_4 + y_6. \tag{1.5}$$

Clearly then, $z = F(\bar{x}) = \binom{7}{11}$. To obtain matrix C, differentiate equations (1.4) and (1.5) with respect to x and y to get

$$A = \begin{pmatrix} 3 & 0 \\ 0 & 2 \\ 2 & 0 \\ 0 & 0 \\ 4 & 0 \\ 0 & 1 \end{pmatrix}, \quad L = \begin{pmatrix} -1 & & & & & \\ & -1 & & & & \\ & & -1 & & & \\ & & 5 & -1 & & \\ & & & & -1 & \\ & & & & 1 & -1 \end{pmatrix},$$

$$B = \begin{pmatrix} 0 & 0 \\ 0 & 0 \end{pmatrix}, \quad M = \begin{pmatrix} 1 & 1 & 0 & 0 & 0 & 0 \\ 0 & 0 & 0 & 1 & 0 & 1 \end{pmatrix}.$$

Note that

$$L^{-1} = \begin{pmatrix} -1 & & & & & \\ & -1 & & & & \\ & & -1 & & & \\ & & -5 & -1 & & \\ & & & & -1 & \\ & & & & -1 & -1 \end{pmatrix},$$

and so it is easy to verify that $B - ML^{-1}A = \begin{pmatrix} 3 & 2 \\ 14 & 1 \end{pmatrix} = J$. ■

The other way to compute (1.2) is

$$J = B - [ML^{-1}]A. \tag{1.6}$$

Here, it is clear that the elements of the matrix must be accessed from bottom to top—hence, L must be stored, in contrast to (1.3). On the other hand, examination of (1.6) yields that the work to evaluate J by this procedure, known as the "reverse mode," is proportional to $m \cdot p$.

In summary, the two modes of automatic differentiation, forward and reverse, can be interpreted as two ways to evaluate (1.2). Forward mode yields the Jacobian in time bounded by $n \cdot \omega(F)$, and the space required is no more than that required to evaluate the function F and store J since forward mode can be implemented "in place." Reverse mode yields the Jacobian in time bounded by $m \cdot \omega(F)$ but requires space to store all the operations; i.e., matrix C must be saved (so that L can be accessed in reverse order).

If structure and sparsity are ignored (or not relevant) and space is not an issue, then time efficiency says to choose forward-mode AD when $n \leq m$ and reverse mode AD otherwise. The subsequent time savings can be significant. Questions of sparsity and structure and reducing the space demands of reverse mode are dealt with in subsequent chapters.

1.2 ▪ The Gradient

The gradient computation is an important special case of the Jacobian computation where $m = 1$. In particular, suppose $f : \mathcal{R}^n \to \mathcal{R}^1$, f is differentiable, and we are interested in computing the gradient: $\nabla f(x) = \left(\frac{\partial f}{\partial x_1}, \frac{\partial f}{\partial x_2}, \ldots, \frac{\partial f}{\partial x_n}\right)^T$. Since $\nabla f^T(x)$ is the Jacobian matrix of function $f(x)$, our previous remarks on Jacobian evaluation by AD hold, with $m = 1$. Specifically, the reverse-mode AD can determine the gradient at an iterate x in time proportional to the time to evaluate the function itself, i.e., $\omega(f)$, whereas the forward-mode AD requires time proportional to $n \cdot \omega(f)$. This is a striking result. A similar contrast exists when comparing reverse-mode AD to finite differences to compute $\nabla f(x)$: the time to determine this gradient by finite differences is proportional to $n \cdot \omega(f)$.

This sharp contrast in computing times is exhibited in practice until space requirements become excessive (e.g., [32]). When this happens, i.e., fast memory is saturated, slower secondary memory may be used, and the theoretical running time bound of $O(\omega(f))$ to evaluate the gradient by the reverse-mode AD may not be observed in practice. In Chapter 4 we show how space demands can often be mitigated, in the presence of structure, and then this striking time bound of $O(\omega(f))$ to obtain the gradient by the reverse mode can still apply for computation-intensive functions.

1.3 ▪ Simple Jacobian and Gradient Examples with ADMAT

In this section we give two examples of how to obtain Jacobian matrices and gradients using ADMAT. Note that these are small unstructured examples, designed to illustrate the basic usage of ADMAT. Subsequent chapters will deal with sparse and structured problems—both very important topics for efficiency!

We assume that the user has installed ADMAT as indicated in Appendix A.

In this first example we illustrate how to obtain a Jacobian matrix, with forward and then reverse modes of AD, using ADMAT. For reasons that will be discussed in subsequent chapters, our code to evaluate Jacobian matrices allows for more generality. In particular, if $F : \mathcal{R}^n \to \mathcal{R}^m$, then ADMAT forward mode allows the user to define an arbitrary matrix V with n rows. ADMAT then computes, using forward-mode AD, the product JV, where $J = J(x) \in \mathcal{R}^{m \times n}$ is the Jacobian matrix. Note, as discussed in Chapter 3, that forward-mode AD ***computes*** JV ***directly*** and does not compute J first. To get the Jacobian matrix, choose $V = I_{n \times n}$. In a similar vein, ADMAT directly computes $J^T W$

for an arbitrary (user-supplied) matrix W with m rows when using the reverse mode. If the Jacobian itself is wanted by reverse mode, then choose $W = I_{m\times m}$. In Chapter 2 we illustrate how a combination of forward and reverse modes can be used in concert to efficiently determine a sparse Jacobian matrix based on the method proposed in [42].

Example 1.2. Compute the Jacobian of the Broyden function by feval.

See DemoJac.m

The Broyden function is derived from a chemical engineering application [17]. Its definition is as follows:

```
function fvec = broyden(x, Extra)
% Evaluate the length of input
   n = length(x);
% Initialize fvec.  It has to be allocated in memory first in ADMAT.
   fvec=zeros(n,1);
   i=2:(n-1);
   fvec(i)= (3-2.*x(i)).*x(i)-x(i-1)-2*x(i+1) + 1;
   fvec(n)= (3-2.*x(n)).*x(n)-x(n-1)+1;
   fvec(1)= (3-2.*x(1)).*x(1)-2*x(2)+1;
```

1. Set the problem size.
   \>> n = 5

2. Set '**broyden**' as the function to be differentiated.
   \>> myfun = ADfun('broyden', n);

   Note that the Broyden function is a vector-valued function that maps R^n to R^n. Thus, the second input argument in **ADfun** is set to n corresponding to the column dimension.

3. Set the independent variable **x** as an "all ones" vector.
   \>> x = ones(n,1)
   ```
   x =
       1
       1
       1
       1
       1
   ```

4. Call **feval** to compute the function value and the Jacobian matrix at x. We omit the input argument (**Extra**) when calling **feval**, given that it is empty.
   \>> [F, J] = feval(myfun, x)
   ```
   F =
        0
       −1
       −1
       −1
        1
   ```

```
J =
    -1  -2   0   0   0
    -1  -1  -2   0   0
     0  -1  -1  -2   0
     0   0  -1  -1  -2
     0   0   0  -1  -1
```

5. Using the forward-mode AD, compute the product $J(x) \times V$, where V is an $n \times 3$ "all ones" matrix. Set the third input argument to "[ ]" since no parameters are stored in Extra.

```
>> V = ones(n,3);
>> options = setopt('forwprod', V);    % set options to forward mode AD
>> [F,JV] = feval(myfun, x, [ ], options)
F =
     0
    -1
    -1
    -1
     1
JV =
    -3  -3  -3
    -4  -4  -4
    -4  -4  -4
    -4  -4  -4
    -2  -2  -2
```

6. Using the reverse-mode AD, compute the product of $J^T(x) \times W$, where W is an $n \times 3$ "all ones" matrix. We pass "[ ]" to the third input argument of feval since no parameters are stored in Extra.

```
>> W = ones(n,3);
>> options = setopt('revprod', W);    % set options to reverse mode AD
>> [F, JTW] = feval(myfun, x, [], options)
F =
     0
    -1
    -1
    -1
     1
JTW =
    -2  -2  -2
    -4  -4  -4
    -4  -4  -4
    -4  -4  -4
    -3  -3  -3
```

■

In the next example, we illustrate how to obtain the gradient of differentiable scalar-valued function $f(x): \mathcal{R}^n \rightarrow \mathcal{R}^1$.

Example 1.3. Compute the gradient of the Brown function.

See DemoGrad.m

This example shows how to compute the gradient of the Brown function in ADMAT in three different ways: default (reverse), forward, and reverse. The definition of the Brown function is as follows:

$$y = \sum_{i=1}^{n-1} \{(x_i^2)^{x_{i+1}^2+1} + (x_{i+1}^2)^{x_i^2+1}\}.$$

A MATLAB function to evaluate the Brown function is given below:

```
function value = brown(x,Extra)
% Evaluate the problem size.
    n = length(x);
% Initialize intermediate variable y.
    y=zeros(n,1);
    i=1:(n-1);
    y(i)=(x(i) .^ 2) .^ (x(i+1) .^ 2+1);
    y(i)=y(i)+(x(i+1) .^ 2) .^ (x(i) .^ 2+1);
    value=sum(y);
```

The following is an illustration of the use of ADMAT with the Brown function:

1. Set the function to be differentiated to be the Brown function.
   >> myfun = ADfun('brown', 1);

   Note: The second input argument in **ADfun**, "1," is a flag indicating a scalar mapping, $f : R^n \to R^1$; more generally, the second argument is set to "m" for a vector-valued function, $F : R^n \to R^m$.

2. Set the dimension of x.
   >> n = 5;

3. Initialize vector x.
   >> x = ones(n,1)
   ```
   x =
       1
       1
       1
       1
       1
   ```

4. Call **feval** to get the function value and the gradient of the Brown function, allowing ADMAT to choose the mode (by default, ADMAT chooses to use reverse mode for computing the gradients).
   >> [f, grad] = feval(myfun, x)
   ```
   f =
       8
   grad =
       4   8   8   8   4
   ```

5. Use the forward mode to compute the gradient of f. In this case the third input argument of **feval** is set to the empty array "[ ]" since no parameters are stored in the input variable **Extra**.
   % Compute the gradient by forward mode
   %

```
% set options to forward mode AD. Input n is the problem size.
>> options = setgradopt('forwprod', n);
% compute gradient
>> [f,grad] = feval(myfun, x, [ ], options)
f =
    8
grad =
    4    8    8    8    4
```

6. Use the reverse mode to compute the gradient g. As in the above case, the input Extra is "[]."

```
% Compute the gradient by reverse mode
%
% set options to reverse mode AD. Input n is the problem size.
>> options = setgradopt('revprod', n);
% compute gradient
>> [f, grad] = feval(myfun, x, [], options)
f =
    8
grad =
    4    8    8    8    4
```

■

1.4 ▪ Computing Second Derivatives and the Hessian Matrix

Many applications require both first and second derivatives. For example, in optimization, consider solving

$$\min_x f(x), \qquad \text{where } f: \mathcal{R}^n \to \mathcal{R}^1,$$

and f is twice continuously differentiable. Many modern algorithms require the triple $\{f(x), \nabla f(x), \nabla^2 f(x)\}$ at each iterate x. The second derivatives of f at point x are represented in the Hessian matrix $\nabla^2 f(x)$, i.e., the real symmetric matrix of second derivatives:

$$\nabla^2 f(x) = \begin{pmatrix} \frac{\partial^2 f}{\partial x_1^2} & \frac{\partial^2 f}{\partial x_1 \partial x_2} & \cdots & \frac{\partial^2 f}{\partial x_1 \partial x_n} \\ \frac{\partial^2 f}{\partial x_2 \partial x_1} & & & \vdots \\ \vdots & & \ddots & \\ \frac{\partial^2 f}{\partial x_n \partial x_1} & \cdots & & \frac{\partial^2 f}{\partial x_n^2} \end{pmatrix}.$$

The question we address here is how to obtain $\nabla^2 f(x)$, via AD, given the code to evaluate $f(x)$. Of course given the code to evaluate $\nabla f(x)$, the approach in Section 1.1 is directly applicable since $\nabla^2 f(x)$ is just the Jacobian of $\nabla f(x)$. In this case, i.e., we have at hand a code to evaluate $\nabla f(x)$, forward mode has the advantage over reverse mode due to the ability to use less space (both forward and reverse will produce the Hessian matrix, i.e., the Jacobian of $\nabla f(x)$, in time proportional to $n \cdot \omega(\nabla f)$). Note that it is probably unlikely that a code "at hand" to evaluate $\nabla f(x)$ runs in time proportional to $\omega(f)$, so this approach may ultimately be slower than the AD approach that works with a code that evaluates $f(x)$. Subsequently, we assume we have at hand a code to evaluate the objective function $f(x)$, not $\nabla f(x)$.

The cost of computing the Hessian matrix by AD, $\omega(\nabla^2 f(x))$, without using any sparsity or structure, assuming a code to evaluate $f(x)$, satisfies

$$\omega(\nabla^2 f) \sim n \cdot \omega(\nabla f) \sim n \cdot \omega(f).$$

Let us examine this situation in more detail. As mentioned above, every nonlinear function can be written as a partially ordered sequence of atomic functions, where each atomic function is in fact a function of one or two previously computed intermediate (or original) variables. We assume each atomic function is twice continuously differentiable and the first and second derivative functions are known. Hence, for a scalar-valued function, we assume that the problem

$$\text{evaluate} \quad f(x)$$

can be equivalently written (see Section 1.1) as

$$\boxed{\begin{array}{ll} \text{solve for } y: & F^E(x,y)=0 \\ \text{solve for output } z: & z-\bar{f}(x,y_1,\dots,y_p)=0, \end{array}} \tag{1.7}$$

where

$$F^E = \begin{pmatrix} F_1 \\ \vdots \\ F_p \end{pmatrix}, \tag{1.8}$$

and each component function F_i determines y_i, in turn, and is a function of one to two variables that have been determined in a previous step, $1,2,\dots,i-1$. Similarly we assume here that $\bar{f}$ is a simple and differentiable unitary or binary function.

So without loss of generality, we assume form (1.7), (1.8). The Jacobian associated with mapping (1.7) can be written

$$J_E = \begin{pmatrix} F_x^E & F_y^E \\ \nabla_x \bar{f}^T & \nabla_y \bar{f}^T \end{pmatrix} \tag{1.9}$$

and

$$\nabla_x f^T = \nabla_x \bar{f}^T - \nabla_y \bar{f}^T (F_y^E)^{-1} F_x^E. \tag{1.10}$$

Matrix J_E corresponds to matrix C in (1.1), partitioned in the same fashion.

Now we make two observations. First, just as the first derivatives of simple functions of one or two variables are well known and easy to compute, so are the second derivatives. Second, in principle the gradient computation defined by (1.9), (1.10) can be evaluated by solving the following system:

$$\boxed{\left.\begin{array}{ll} \text{solve for } (x,y) \text{ and differentiate w.r.t. } x,\, y: & F^E(x,y)=0 \\ \text{solve for } w: & (F_y^E)^T w + \nabla_y \bar{f} = 0 \\ \text{solve for output(gradient) } z: & (F_x^E)^T w + \nabla_x \bar{f} - z = 0 \end{array}\right\}} \tag{1.11}$$

Note that all systems in (1.11) are triangular and therefore can be solved by a substitution process. Assuming all atomic functions are twice continuously differentiable, system (1.11) can be differentiated to yield a Jacobian with respect to x, y, w:

$$H_E = \begin{pmatrix} F_x^E & F_y^E & 0 \\ (F_{yx}^E)^T w + \nabla_{yx}^2 \bar{f} & (F_{yy}^E)^T w + \nabla_{yy}^2 \bar{f} & (F_y^E)^T \\ (F_{xx}^E)^T w + \nabla_{xx}^2 \bar{f} & (F_{xy}^E)^T w + \nabla_{xy}^2 \bar{f} & (F_x^E)^T \end{pmatrix}. \tag{1.12}$$

All terms in (1.12) are first or second derivatives of the atomic functions that define our objective function. Hence, they can be easily computed by AD using a code to evaluate the objective function. System (1.12) was obtained by differentiating an expression for the gradient term; therefore, we can expect it is related to our desired Hessian matrix. We will illustrate the relationship below. First, let us simplify (1.12) by defining a function

$$g(x,y,w) = \bar{f} + w^T F^E(x,y). \tag{1.13}$$

Then (1.12) is simply

$$H_E = \begin{pmatrix} F_x^E & F_y^E & 0 \\ \nabla_{yx}^2 g & \nabla_{yy}^2 g & (F_y^E)^T \\ \nabla_{xx}^2 g & \nabla_{xy}^2 g & (F_x^E)^T \end{pmatrix}. \tag{1.14}$$

Here is a recipe for computing the extended Hessian H_E:

1. Evaluate f and compute, by differentiation, the extended Jacobian, (F_x^E, F_y^E).
2. Solve the triangular system for w, $(F_y^E)^T w + \nabla_y \bar{f} = 0$.
3. Twice differentiate the function $g(x,y,w) = \bar{f} + w^T F^E(x,y)$ with respect to (x,y). Note that this function is simply the weighted sum of many atomic functions, i.e., simple (and easy to differentiate) functions.

The computing cost of obtaining (1.14) is proportional to the cost of computing the objective function, i.e., $\omega(H_E) \sim \omega(f) \sim p$. This proportionality relationship follows from the observation that the cost of evaluating the first and second derivatives of any of the simple functions of one to two variables is a constant, and there are p such terms in f. Note that if we interchange columns 1 and 3, then we obtain symmetry:

$$H_E^S = \begin{pmatrix} 0 & F_y^E & F_x^E \\ (F_y^E)^T & \nabla_{yy}^2 g & \nabla_{yx}^2 g \\ (F_x^E)^T & \nabla_{xy}^2 g & \nabla_{xx}^2 g \end{pmatrix}. \tag{1.15}$$

Symmetry of the matrices on the diagonal in (1.15) will follow if the atomic functions are twice continuously differentiable.

But where is the Hessian matrix $H \triangleq \nabla^2 f(x)$?

The answer is that H can be extracted from H_E by a Schur-complement computation.

Specifically, if we partition H_E

$$J^E = \left(\begin{array}{c|cc} F_x^E & F_y^E & 0 \\ \nabla_{yx}^2 g & \nabla_{yy}^2 g & (F_y^E)^T \\ \hline \nabla_{xx}^2 g & \nabla_{xy}^2 g & (F_x^E)^T \end{array}\right) = \begin{pmatrix} A & L \\ B & M \end{pmatrix}, \tag{1.16}$$

then our desired matrix $H = B - ML^{-1}A$. Note that matrix L is a mix of a nonsingular lower triangular matrix, F_y^E, and a rectangular upper triangular matrix, $(F_y^E)^T$, and so either $L^{-1}A$ or ML^{-1} can be solved by a substitution process. Moreover, since the number of columns in matrix A equals the number of rows of $M(=n)$ and space requirements are similar (matrix F_y^E must be stored since access in both directions is required), there is no obvious advantage in the order of computation of $ML^{-1}A$ in this case (unlike the general situation discussed in Section 1.1).

The work required by the above procedure, given the n columns of A (or n rows of M), is $\omega(H) \sim n \cdot \omega(f) \sim n \cdot p$, and space is proportional to p.

Example 1.4. Let $f(x) = x_1^2 + 2x_2^2 + 4x_1x_2$ and let the current point be $\bar{x}^T = (-1, 1)$. Clearly then, by substitution, $z \triangleq f(\bar{x}) = -1$, $\nabla f^T(\bar{x}) = (2, 0)$, $H = \left(\begin{smallmatrix} 2 & 4 \\ 4 & 4 \end{smallmatrix}\right)$. The function F^E can be defined by a sequence:

$$\begin{cases} y_1 = x_1^2,\ y_2 = x_2^2,\ y_3 = 2y_2,\ y_4 = x_1x_2, \\ y_5 = 4y_4,\ y_6 = y_1 + y_3,\ z = y_5 + y_6. \end{cases} \tag{1.17}$$

The current value of the vector y of intermediate variables, denoted by $\bar{y}$, using (1.17) is $\bar{y} = (1, 1, 2, -1, -4, 3, -1)^T$. The corresponding extended Jacobian at $(\bar{x}, \bar{y})$ is

$$J_E = \left(\begin{array}{cc|cccccc} -2 & 0 & -1 & & & & & \\ & 2 & & -1 & & & & \\ & & & 2 & -1 & & & \\ 1 & -1 & & & & -1 & & \\ & & & & & 4 & -1 & \\ & & 1 & & 1 & & & -1 \\ \hline & & & & & & 1 & 1 \end{array}\right) = \begin{pmatrix} F_x^E & F_y^E \\ \nabla_x \bar{f}^T & \nabla_y \bar{f}^T \end{pmatrix}. \tag{1.18}$$

The gradient, as indicated by (1.10), is $\nabla_x f^T = \nabla_x \bar{f}^T - \nabla_y \bar{f}^T (F_y^E)^{-1} F_x^E$. Note that

$$(F_y^E)^{-1} = \begin{pmatrix} -1 & & & & & \\ & -1 & & & & \\ & -2 & -1 & & & \\ & & & -1 & & \\ & & & -4 & -1 & \\ -1 & -2 & -1 & & & -1 \end{pmatrix}. \tag{1.19}$$

It is straightforward to verify, using $\nabla_x f^T = \nabla_x \bar{f}^T - \nabla_y \bar{f}^T (F_y^E)^{-1} F_x^E$, that $\nabla_x f^T(\bar{x}) = (2,0)$. Now using (1.13),

$$g(x,y,w) = \bar{f} + w^T F^E(x,y)$$
$$= (y_5 + y_6) + (1,2,1,4,1,1) \begin{pmatrix} x_1^2 - y_1 \\ x_2^2 - y_2 \\ 2y_2 - y_3 \\ x_1 x_2 - y_4 \\ 4y_4 - y_5 \\ y_1 + y_3 - y_6 \end{pmatrix}. \tag{1.20}$$

And so, after multiplying through,[2]

$$\begin{aligned} g(x,y) = &(y_5 + y_6) + (x_1^2 - y_1) + 2(x_2^2 - y_2) + (2y_2 - y_3) \\ &+ 4(x_1 x_2 - y_4) + (4y_4 - y_5) + (y_1 + y_3 - y_6). \end{aligned}$$

Clearly, in this case

$$\left\{ \begin{array}{l} \nabla_{yx}^2 g = 0_{p\times n},\ \nabla_{xy}^2 g = 0_{n\times p},\ \nabla_{yy}^2 g = 0_{p\times p} \\ \nabla_{xx}^2 g = \left(\begin{smallmatrix} 2 & 4 \\ 4 & 4 \end{smallmatrix}\right) \end{array} \right. ,$$

and so all blocks of matrix H_E in (1.16) have been determined, and using $H = B - ML^{-1}A$, we obtain

$$H = \begin{pmatrix} 2 & 4 \\ 4 & 4 \end{pmatrix} - 0_{2\times 2} = \begin{pmatrix} 2 & 4 \\ 4 & 4 \end{pmatrix},$$

which is the desired result. ■

To follow are two examples of the use of ADMAT to obtain Hessian information.

Example 1.5. Compute the first- and second-order derivatives of $y = x^2$.
See Demo2nd1.m

1. Define a derivtapeH object with value 3.
```
>> x = derivtapeH(3,1,1)
val =
     3
varcount =
     1
val =
     1
varcount =
     2
```

2. Compute $y = x^2$.
```
>> y = x ^ 2
val =
     9
varcount =
     3
```

[2]We now write $g(x,y)$ since w is now determined and fixed.

```
val =
     6
varcount =
     6
```

3. Get the value of y.

```
>> yval = getval(y)
yval =
     9
```

4. Get the first-order derivative of y.

```
>> y1d = getydot(y)
y1d =
     6
```

5. Get the second-order derivative of y.

```
>> y2d = parsetape(1)
y2d =
     2
```

■

Example 1.6. Compute the gradient and Hessian of the Brown function.
See Demo2nd2.m

1. Set the problem size.

```
>> n = 5        % problem size
```

2.

```
>> x = ones(n,1)
x =
     1
     1
     1
     1
     1
```

3. Define a derivtapeH object.

```
>> x = derivtapeH(x,1,eye(n))
val -
     1
     1
     1
     1
     1
varcount =
     7
val =
     1  0  0  0  0
     0  1  0  0  0
     0  0  1  0  0
     0  0  0  1  0
     0  0  0  0  1
varcount =
     8
```

4. Call the Brown function.

```
>> y = brown(x)
val =
    8
varcount =
    95
val =
    4  8  8  8  4
varcount =
    96
```

5. Get the value of y.

```
>> yval = getval(y)
yval =
    8
```

6. Get the gradient of the Brown function at x.

```
>> grad = getydot(y)
grad =
    4  8  8  8  4
```

7. Get the Hessian of the Brown function at x.

```
>> H = parsetape(eye(n))
H =
   12   8   0   0   0
    8  24   8   0   0
    0   8  24   8   0
    0   0   8  24   8
    0   0   0   8  12
```

■

Chapter 2

Products and Sparse Problems

AD technology enables the computation of Jacobian-matrix (and Hessian-matrix) products directly, without first requiring the determination of the Jacobian (Hessian) matrix itself. For example, if the product JV is required where J is the m-by-n Jacobian of the differentiable mapping $F : \mathcal{R}^n \rightarrow \mathcal{R}^m$ and V is an n-by-t_V matrix, then this product can be determined directly by the forward mode of AD. Specifically, (1.2) becomes

$$JV = BV - M[L^{-1}AV]. \tag{2.1}$$

The work to compute JV in this way is then $\omega(JV) \sim t_V \cdot \omega(F)$. This contrasts with work required to compute J first and then multiply to get JV, which, in the unstructured, non-sparse case, is proportional to $n \cdot \omega(F)$ plus the work to perform the matrix multiplication.

Similarly, if W is an m-by-t_W matrix, the product W^TJ can be computed by the reverse mode, i.e., modify (1.8) to set

$$J = W^T B - [W^T ML^{-1}]A, \tag{2.2}$$

and the work required is $\omega(W^TJ) \sim t_W \cdot \omega(F)$.

Clearly the savings can be substantial when the number of columns of V (or W) is small relative to the column (or row) dimension. The question of whether to first compute the product or first compute J and then multiply is relatively straightforward when J (or H) is dense and unstructured and the matrix V (or W) is known in advance. If there is sparsity or structure, the trade-offs become more subtle—we address these situations subsequently. In addition, if the number of columns in V (or W) is not known in advance (say in an iterative linear system approach), then an adjustable approach may be desired.

One application of product determination occurs in the constrained optimization. Consider the problem

$$\min\{f(x) : Ax = b\}, \tag{2.3}$$

where f is twice continuously differentiable and A is an $m \times n$ matrix with $m < n$. One approach to solve this problem is to determine a basis Z for the null space of A and then work with the reduced gradient and Hessian matrices, $Z^T\nabla f(x)$ and $Z^T\nabla^2 f(x)Z$, respectively. AD can be used to determine these quantities. However, there is an efficiency choice for the reduced Hessian matrix computation. Specifically, it is possible to use AD to determine $\nabla^2 f(x)$ and then multiply to obtain $Z^T \nabla^2 f(x) Z$ or instead use AD to determine $\nabla^2 f(x)Z$ directly and then multiple this result by Z^T. Which is the better approach?

The answer to this question depends on the possible presence of sparsity (discussed subsequently) in $\nabla^2 f(x)$. If this matrix is dense (i.e., little or no sparsity), then the latter technique is more efficient (i.e., determine $\nabla^2 f Z$ by AD and then multiply the result by Z^T). This is because the work for this approach is proportional to $(m-n)\cdot\omega(f)$, assuming A is of full row rank, which is clearly better than $m \cdot \omega(f)$. However, as we indicate below, the comparison is more subtle when $\nabla^2 f$ is sparse.

2.1 ▪ Sparsity

There is a useful connection between efficiently determining the nonzero elements of a sparse Jacobian (Hessian) matrix and determining Jacobian-matrix (Hessian-matrix) products. In particular, an effective way to determine the nonzero elements of a sparse Jacobian matrix is to determine a thin matrix V and/or a thin matrix W, determine JV (by forward-mode AD) and/or determine $W^T J$ (by reverse-mode AD), and then recover the nonzeros of J from these products.

The question is how to find such matrices V, W such that they are "thin" and all nonzeros can be effectively recovered from these products. Let us consider three examples first before discussing the general sparse situation.

Suppose $F : \mathcal{R}^n \rightarrow \mathcal{R}^n$ is differentiable and the Jacobian of F has the following sparse structure (illustrated for the case $n = 5$):

$$J = \begin{pmatrix} X_1 & & & & \\ X_2 & Y_1 & & & \\ X_3 & & Y_2 & & \\ X_4 & & & Y_3 & \\ X_5 & & & & Y_4 \end{pmatrix}, \tag{2.4}$$

where X and Y indicate nonzero values. Suppose $V = \left(\begin{smallmatrix} 1 & 0 & 0 & 0 & 0 \\ 0 & 1 & 1 & 1 & 1 \end{smallmatrix}\right)^T$. The computation of JV yields

$$JV = \begin{pmatrix} X_1 & 0 \\ X_2 & Y_1 \\ X_3 & Y_2 \\ X_4 & Y_3 \\ X_5 & Y_4 \end{pmatrix}. \tag{2.5}$$

Clearly all nonzeros of J are recovered from (2.5). The X-values are recovered in the first column of JV, and the Y-values are recovered in the second column of JV. Matrix V has only two columns regardless of the size of n provided the structure in (2.4) holds.

Suppose that the structure of J is transposed, i.e.,

$$J = \begin{pmatrix} X_1 & X_2 & X_3 & X_4 & X_5 \\ & Y_1 & & & \\ & & Y_2 & & \\ & & & Y_3 & \\ & & & & Y_4 \end{pmatrix}. \tag{2.6}$$

In this case choose $W = \begin{pmatrix} 1&0&0&0&0 \\ 0&1&1&1&1 \end{pmatrix}^T$ and compute

$$W^T J = \begin{pmatrix} X_1 & X_2 & X_3 & X_4 & X_5 \\ 0 & Y_1 & Y_2 & Y_3 & Y_4 \end{pmatrix}. \tag{2.7}$$

Clearly all nonzero elements of the Jacobian matrix are recovered (by the reverse-mode in this case) using a matrix W with just two columns (regardless of the value of n).

It turns out that because of the dense first row in (2.6), the forward mode of AD would do poorly in this example; similarly, the reverse mode does poorly in example (2.4) due to the dense first column. Now consider the following example:

$$J = \begin{pmatrix} X_1 & X_2 & X_3 & X_4 & X_5 \\ Z_1 & Y_1 & & & \\ Z_2 & & Y_2 & & \\ Z_3 & & & Y_3 & \\ Z_4 & & & & Y_4 \end{pmatrix}. \tag{2.8}$$

The dense first row is disadvantageous for forward mode, which requires work proportional to $n \cdot \omega(F)$. Similarly, because of the dense first column, reverse-mode AD will determine J in time proportional to $n \cdot \omega(F)$. So neither forward mode nor reverse mode is efficient in this case. However, a suitable combination of forward and reverse modes allows J to be determined in time proportional to $3 \cdot \omega(F)$. Specifically, let

$$V = \begin{pmatrix} 1 \\ 0 \\ 0 \\ 0 \\ 0 \end{pmatrix} \quad \text{and} \quad W = \begin{pmatrix} 1 & 0 \\ 0 & 1 \\ 0 & 1 \\ 0 & 1 \\ 0 & 1 \end{pmatrix}.$$

Then the forward mode determines

$$JV = \begin{pmatrix} X_1 \\ Z_1 \\ Z_2 \\ Z_3 \\ Z_4 \end{pmatrix},$$

and the reverse mode determines

$$W^T J = \begin{pmatrix} X_1 & X_2 & X_3 & X_4 & X_5 \\ - & Y_1 & Y_2 & Y_3 & Y_4 \end{pmatrix},$$

and clearly all nonzeros of matrix J are determined. Note that in this case the (2,1)-location of the product $W^T J$ is of no use.

2.2 ▪ Partition Problems and Graph Theory

Our basic task is to efficiently determine thin matrices V, W so that the nonzero elements of J can be readily extracted from the information (W^TJ, JV). The pair of matrices (W^TJ, JV) is obtained from the application of both modes of automatic differentiation: matrix W^TJ is computed by the reverse mode, and the forward mode determines JV. The purpose of this section is to more rigorously formulate the question of determining suitable matrices V, W, first in the language of "partitions" and then using graph theoretic concepts. This approach is due to [42].

A *bi-partition* of a matrix J is a pair (G_R, G_C) where G_R is a row partition of a *subset* of the rows of J and G_C is a column partition of a *subset* of the columns of J.

In an example illustrated from [42], the use of a bi-partition dramatically decreases the amount of work required to determine J as illustrated in Figure 2.1. Specifically, the total amount of work required in this case is proportional to $3 \cdot \omega(F)$. To see this, define $V = (e_1, e_2 + e_3 + e_4 + e_5)$ and $W = (e_1)$, where we follow the usual convention of representing the ith column of the identity matrix with e_i. Clearly elements $\square, \triangle$ are directly determined from the product JV; elements $\triangle$ are directly determined from the product W^TJ.

The basic idea is to partition the rows into a set of groups G_R and the columns into a set of groups G_C, with $|G_R| + |G_C|$ as small as possible. $|G_C|$ and $|G_R|$ represent the number of groups in G_C and G_R, respectively, such that every nonzero element of J can be directly determined from *either* a group in G_R or a group in G_C.

A bi-partition (G_R, G_C) of a matrix J is *consistent with direct determination* if for every nonzero $\alpha_{i,j}$ of J, either column j is in a group of G_C, which has no other column having a nonzero in row i, or row i is in a group of G_R, which has no other rows having a nonzero in column j.

Clearly, given a bi-partition (G_R, G_C) *consistent with direct determination*, we can trivially construct matrices $W \in \Re^{m \times |G_R|}$, $V \in \Re^{n \times |G_C|}$ such that J can be directly determined from (W^TJ, JV).

If we relax the restriction that each nonzero element of J can be determined directly, then it is possible that the work required to evaluate the nonzeros of J can be further reduced. For example, we could allow for a "substitution" process when recovering the nonzeros of J from the pair (W^TJ, JV). Figures 2.1 and 2.2 illustrate that a substitution method can win over direct determination: Figure 2.1 corresponds to direct determination, and Figure 2.2 corresponds to determination using substitution.

In both cases, elements labeled $\square$ are computed from the column groupings, i.e., calculated using the product JV; elements labeled $\triangle$ are calculated from the row groupings, i.e., calculated using the product W^TJ. The matrix in Figure 2.1 indicates that we can choose G_C with $|G_C| = 2$ and G_R with $|G_R| = 1$ and determine all elements directly. That is, choose $V = (e_1 + e_7, e_4)$; choose $W = (e_1 + e_4 + e_7)$. Therefore, in this case the work to compute J satisfies $\omega(J) \sim 3 \cdot \omega(F)$. Note that some elements can be determined twice, e.g., J_{11}. However, the matrix in Figure 2.2 shows how to obtain the nonzeros of J, using substitution, in work proportional to $2 \cdot \omega(F)$. Let $V = W = (e_1 + e_4 + e_7)$. Let p_1 be

$$\begin{pmatrix}
\Box_{11} & \triangle_{12} & \triangle_{13} & & & & & & \\
\Box_{21} & & & & & & & & \\
\Box_{31} & & & & & & & & \\
\Box_{41} & & & \Box_{44} & \triangle_{45} & \triangle_{46} & & & \\
 & & & \Box_{54} & & & & & \\
 & & & \Box_{64} & & & & & \\
 & & & \Box_{74} & & & \Box_{77} & \triangle_{78} & \triangle_{79} \\
 & & & & & & \Box_{87} & & \\
 & & & & & & \Box_{97} & & \\
 & & & & & & \Box_{10,7} & &
\end{pmatrix}$$

Figure 2.1. *Optimal partition for direct method.*

$$\begin{pmatrix}
\Box_{11} & \triangle_{12} & \triangle_{13} & & & & & & \\
\Box_{21} & & & & & & & & \\
\Box_{31} & & & & & & & & \\
\Box_{41} & & & \triangle_{44} & \triangle_{45} & \triangle_{46} & & & \\
 & & & \Box_{54} & & & & & \\
 & & & \Box_{64} & & & & & \\
 & & & \Box_{74} & & & \triangle_{77} & \triangle_{78} & \triangle_{79} \\
 & & & & & & \Box_{87} & & \\
 & & & & & & \Box_{97} & & \\
 & & & & & & \Box_{10,7} & &
\end{pmatrix}$$

Figure 2.2. *Optimal partition for substitution method.*

the (forward) computed vector $p_1 = JV$; let p_2^T be the (reverse) computed row vector $p_2^T = W^T J$. Then

$$p_1 = \begin{pmatrix} \Box_{11} \\ \Box_{21} \\ \Box_{31} \\ \Box_{41} + \triangle_{44} \\ \Box_{54} \\ \Box_{64} \\ \Box_{74} + \triangle_{77} \\ \Box_{87} \\ \Box_{97} \\ \Box_{10,7} \end{pmatrix},$$

$$p_2^T = \left(\Box_{11} + \Box_{41}, \triangle_{12}, \triangle_{13}, \triangle_{44} + \Box_{74}, \triangle_{45}, \triangle_{46}, \triangle_{77}, \triangle_{78}, \triangle_{79}\right).$$

Most of the nonzero elements are determined directly (i.e., no conflict). The remaining elements can be resolved:

$$\Box_{74} = p_1(7) - p_2(7); \quad \triangle_{44} = p_2(4) - \Box_{74}; \quad \Box_{41} = p_1(4) - \triangle_{44}.$$

It is easy to extend this example so that the difference between the number of groups needed, by substitution and direct determination, increases with the dimension of the matrix. For example, a block generalization is illustrated in Figure 2.3: if we assume $l > 2w$, it is straightforward to verify that in the optimal partition the number of groups needed for direct determination will be $3w$, and determination by substitution requires

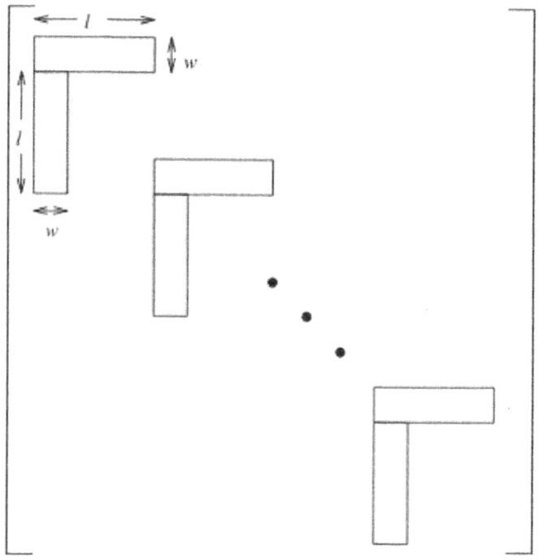

Figure 2.3. *Block example.*

$2w$ groups. A bi-partition (G_R, G_C) of a matrix J is *consistent with determination by substitution*, if there exists an ordering π on elements $\alpha_{i,j}$, such that for every nonzero $J_{i,j}$ of J, either column j is in a group where all nonzeros in row i, from other columns in the group, are ordered lower than $\alpha_{i,j}$, or row i is in a group where all the nonzeros in column j, from other rows in the group, are ordered lower than $\alpha_{i,j}$.

In the usual way we can construct a matrix V from the column grouping G_C and a matrix W from the row grouping G_R. For example, to construct the columns of V, associate with each group in G_C a Boolean vector, with unit entries indicating membership of the corresponding columns. We can now state our main problem(s) more precisely:

The bi-partition problem (direct): Given a matrix J, obtain a bi-partition (G_R, G_C) consistent with direct determination, such that total number of groups, $|G_R| + |G_C|$, is minimized.

The bi-partition problem (substitution): Given a matrix J, obtain a bi-partition (G_R, G_C) consistent with determination by substitution, such that total number of groups, $|G_R| + |G_C|$, is minimized.

Bi-coloring

The two combinatorial problems we face, corresponding to direct determination and determination by substitution, can both be approached in the following way [38]. First, permute and partition the structure of J: $\tilde{J} = P \cdot J \cdot Q = [J_C | J_R]$, as indicated in Figure 2.4. The construction of this partition is crucial; however, we postpone that discussion until after we illustrate its utility. Assume $P = Q = I$ and $J = [J_C | J_R]$. Second, define appropriate intersection graphs $\mathcal{G}_C^I, \mathcal{G}_R^I$ based on the partition $[J_C | J_R]$; a coloring of $\mathcal{G}_C^I$ yields a partition of a subset of the columns, G_C, which defines matrix V. Matrix W is defined by a partition of a subset of rows, G_R, which is given by a coloring of $\mathcal{G}_R^I$. We call this double-coloring approach *bi-coloring*. The difference between the direct and

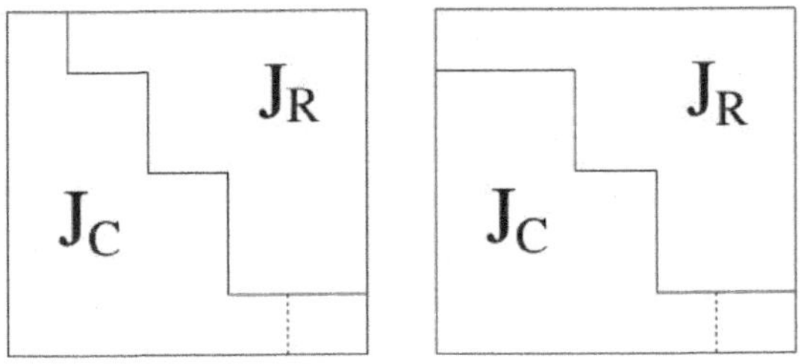

Figure 2.4. *Possible partitions of the matrix* $\tilde{J} = P \cdot J \cdot Q$.

the substitution cases is in how the intersection graphs, $\mathcal{G}_C^I, \mathcal{G}_R^I$, are defined and how the nonzeros of J are extracted from the pair $(W^T J, JV)$.

Direct Determination

In the direct case, the intersection graph $\mathcal{G}_C^I$ is defined to be $\mathcal{G}_C^I = (\mathcal{V}_C^I, \mathcal{E}_C^I)$, where

- vertex $j \in \mathcal{V}_C^I$ if $nnz^3(\text{column } j \cap J_C) \neq 0$;
- $(r,s) \in \mathcal{E}_C^I$ if $r \in \mathcal{V}_C^I$, $s \in \mathcal{V}_C^I$, $\exists k$ such that $J_{kr} \neq 0, J_{ks} \neq 0$ and *either* $(k,r) \in J_C$ *or* $(k,s) \in J_C$.

The key point in the construction of graph $\mathcal{G}_C^I$ and why $\mathcal{G}_C^I$ is distinguished from the usual column intersection graph is that columns r and s are said to intersect if and only if their nonzero locations overlap, in part, in J_C: i.e., columns r and s intersect if $J_{kr} \cdot J_{ks} \neq 0$ and *either* $(k,r) \in J_C$ *or* $(k,s) \in J_C$ for some k.

The "transpose" of the procedure above is used to define $\mathcal{G}_R^I = (\mathcal{V}_R^I, \mathcal{E}_R^I)$. Specifically, $\mathcal{G}_R^I = (\mathcal{V}_R^I, \mathcal{E}_R^I)$, where

- vertex $i \in \mathcal{V}_R^I$ if $nnz(\text{row } i \cap J_R) \neq 0$;
- $(r,s) \in \mathcal{E}_R^I$ if $r \in \mathcal{V}_R^I$, $s \in \mathcal{V}_R^I$, $\exists k$ such that $J_{rk} \neq 0, J_{sk} \neq 0$ and *either* $(r,k) \in J_R$ *or* $(s,k) \in J_R$.

In this case, the reason why graph $\mathcal{G}_R^I$ is distinguished from the usual row intersection graph is that rows r and s are said to intersect if and only if their nonzero locations overlap; more specifically, in part, in J_R: rows r and s intersect if $J_{rk} \cdot J_{sk} \neq 0$ and *either* $(r,k) \in J_R$ *or* $(s,k) \in J_R$ for some k.

The bi-partition (G_R, G_C), induced by coloring of graphs $\mathcal{G}_R^I$ and $\mathcal{G}_C^I$, is consistent with direct determination of J. To see this, consider a nonzero element $(i,j) \in J_C$: column j is in a group of G_C (corresponding to a color) with the property that no other column in G_C has a nonzero in row i. Hence, element (i,j) can be directly determined. Analogously, consider a nonzero element $(r,s) \in J_R$: row r will be in a group of G_R (corresponding to a color) with the property that no other row in G_R has a nonzero in column s. Hence, element (r,s) can be directly determined. Since every nonzero of J is covered, the result follows.

[3] If M is a matrix or a vector, then "$nnz(M)$" is the number of nonzeros in M.

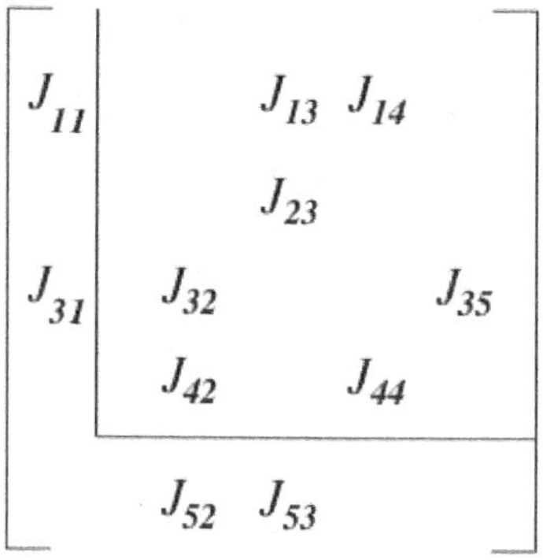

Figure 2.5. *Example partition.*

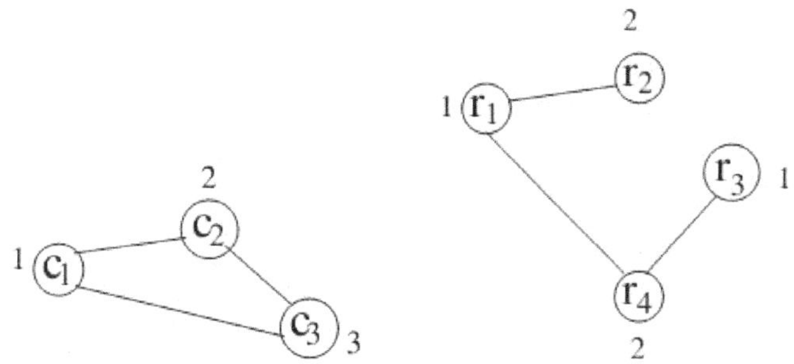

Figure 2.6. *Graphs $\mathcal{G}_C^I$ and $\mathcal{G}_R^I$ (direct approach).*

Example 2.1. Consider the example Jacobian matrix structure as shown in Figure 2.5 with the partition (J_C,J_R) shown. ■

The graphs $\mathcal{G}_C^I$ and $\mathcal{G}_R^I$, formed by the algorithm outlined above, are given in Figure 2.6. Coloring $\mathcal{G}_C^I$ requires three colors, while $\mathcal{G}_R^I$ can be colored in two. Boolean matrices V and W can be formed in the usual way: each column corresponds to a group (or color), and unit entries indicate column (or row) membership in that group:

$$V = \begin{pmatrix} 1 & 0 & 0 \\ 0 & 1 & 0 \\ 0 & 0 & 1 \\ 0 & 0 & 0 \\ 0 & 0 & 0 \end{pmatrix}, \qquad JV = \begin{pmatrix} J_{11} & 0 & \times \\ 0 & 0 & \times \\ J_{31} & \times & \times \\ 0 & \times & 0 \\ 0 & J_{52} & J_{53} \end{pmatrix},$$

$$W = \begin{pmatrix} 1 & 0 \\ 0 & 1 \\ 1 & 0 \\ 0 & 1 \\ 0 & 0 \end{pmatrix}, \qquad W^T J = \begin{pmatrix} \times & J_{32} & J_{13} & J_{14} & J_{35} \\ \times & J_{42} & J_{23} & J_{44} & 0 \end{pmatrix}.$$

Clearly all nonzero entries of J can be identified in either JV or $W^T J$.

Determination by Substitution

The basic advantage of determination by substitution in conjunction with partition $J = [J_C | J_R]$ is that sparse intersection graphs $\mathcal{G}_C^I, \mathcal{G}_R^I$ can be used. Sparser intersection graphs mean thinner matrices V, W, which, in turn, result in reduced cost.

In the substitution case the intersection graph $\mathcal{G}_C^I$ is defined as $\mathcal{G}_C^I = (\mathcal{V}_C^I, \mathcal{E}_C^I)$, where

- vertex $j \in \mathcal{V}_C^I$ if $nnz(\text{column } j \cap J_C) \neq 0$;

- $(r,s) \in \mathcal{E}_C^I$ if $r \in \mathcal{V}_C^I$, $s \in \mathcal{V}_C^I$, $\exists k$ such that $J_{kr} \neq 0$, $J_{ks} \neq 0$ and **both** $(k,r) \in J_C$, $(k,s) \in J_C$.

Note that the intersection graph $\mathcal{G}_C^I = (\mathcal{V}_C^I, \mathcal{E}_C^I)$ captures the notion of two columns intersecting if there is overlap in nonzero structure in J_C: columns r and s intersect if $J_{kr} \cdot J_{ks} \neq 0$, **both** $(k,r) \in J_C$ and $(k,s) \in J_C$ for some k. It is easy to see that $\mathcal{E}_C^I$ is a subset of the set of edges used in the direct determination case.

The "transpose" of the procedure above is used to define $\mathcal{G}_R^I = (\mathcal{V}_R^I, \mathcal{E}_R^I)$. Specifically, $\mathcal{G}_R^I = (\mathcal{V}_R^I, \mathcal{E}_R^I)$, where

- vertex $i \in \mathcal{V}_R^I$ if $nnz(\text{row } i \cap J_R) \neq 0$;

- $(r,s) \in \mathcal{E}_R^I$ if $r \in \mathcal{V}_R^I$, $s \in \mathcal{V}_R^I$, $\exists k$ such that $J_{rk} \neq 0$, $J_{sk} \neq 0$ and **both** $(r,k) \in J_R$, $(s,k) \in J_R$.

The intersection graph $\mathcal{G}_R^I = (\mathcal{V}_R^I, \mathcal{E}_R^I)$ captures the notion of two rows intersecting if there is overlap in nonzero structure in J_R: rows r and s intersect if $J_{rk} \cdot J_{sk} \neq 0$ and $(r,k) \in J_R$ **and** $(s,k) \in J_R$ for some k. It is easy to see that $\mathcal{E}_R^I$ is a subset of the set of edges used in direct determination.

All the elements of J can be determined from $(W^T J, JV)$ by a substitution process. This is evident from the illustrations in Figure 2.7.

Figure 2.7 illustrates two of four possible nontrivial types of partitions. In both cases it is clear that nonzero elements in section 1 can be solved or will be in different groups—because of the construction process. Nonzero elements in 2 either can be determined directly or will depend on elements in section 1. But elements in section 1 are already determined (directly), and so, by substitution, elements in 2 can be determined after 1. Elements in section 3 can then be determined, depending only on elements in 1 and 2 and so on until the entire matrix is resolved.

Example 2.2. Consider again the example Jacobian matrix structure as shown in Figure 2.5. Column and row intersection graphs corresponding to substitution are given in Figure 2.8. Note that $\mathcal{G}_C$ is disconnected and requires two colors; $\mathcal{G}_R$ is a simple chain and also requires two colors. ■

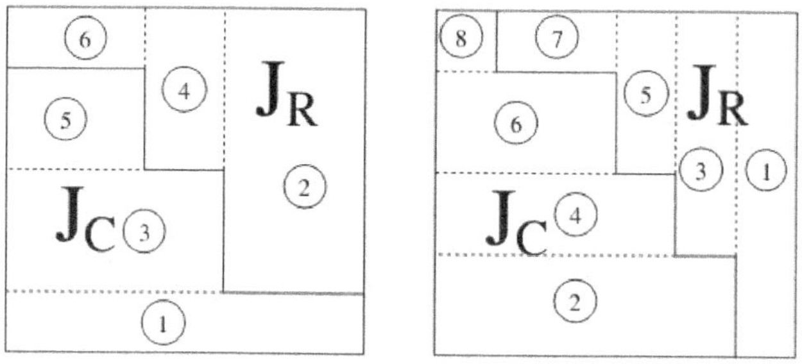

Figure 2.7. *Substitution orderings.*

Figure 2.8. *Graphs $\mathcal{G}_C$ and $\mathcal{G}_R$ for substitution process.*

The coloring of $\mathcal{G}_C$ and $\mathcal{G}_C$ leads to the following matrices V, W and the resulting computation of JV, W^TJ:

$$V = \begin{pmatrix} 1 & 0 \\ 0 & 1 \\ 1 & 0 \\ 0 & 0 \\ 0 & 0 \end{pmatrix}, \qquad JV = \begin{pmatrix} J_{11}+J_{13} & 0 \\ \times & 0 \\ J_{31} & \times \\ 0 & \times \\ J_{53} & J_{52} \end{pmatrix},$$

$$W = \begin{pmatrix} 1 & 0 \\ 0 & 1 \\ 1 & 0 \\ 0 & 1 \\ 0 & 0 \end{pmatrix}, \qquad W^TJ = \begin{pmatrix} \times & J_{32} & J_{13} & J_{14} & J_{35} \\ \times & J_{42} & J_{23} & J_{44} & 0 \end{pmatrix}.$$

It is now easy to verify that all nonzeros of J can be determined via substitution.

2.3 ▪ How to Partition J

We now consider the problem of obtaining a useful partition $[J_C | J_R]$ and the corresponding permutation matrices P, Q, as illustrated in Figure 2.4. A simple heuristic proposed in [38] is based on the knowledge that the subsequent step, in both the direct and the substitution method, is to color intersection graphs based on this partition.

Algorithm **MNCO** builds the partition J_C from the bottom up and partition J_R from right to left. At the kth major iteration, either a new row is added to J_C or a new column is added to J_R: the choice depends on considering a lower bound effectčž

$$\rho(J_R^T) + max(\rho(J_C), nnz(r)) < \rho(J_C) + max(\rho(J_R^T), nnz(c)),$$

where $\rho(A)$ is the maximum number of nonzeros in any row of matrix A, r is a row under consideration to be added to J_C, and c is a column under consideration to be added to J_R. Hence, the number of colors needed to color $\mathcal{G}_C^I$ is bounded below by $\rho(J_C)$, and the number of colors needed to color $\mathcal{G}_R^I$ is bounded below by $\rho(J_R^T)$.

In algorithm **MNCO**, matrix $M = J(R, C)$ is the submatrix of J defined by row indices R and column indices C: M consists of rows and columns of J not yet assigned to either J_C or J_R. The **MNCO** algorithm can be described as follows.

ALGORITHM 2.1.
Minimum Nonzero Counter Ordering (MNCO)

1. Initialize $R = (1 : m)$, $C = (1 : n)$, $M = J(R, C)$;
2. Find $r \in R$ with fewest nonzeros in M;
3. Find $c \in C$ with fewest nonzeros in M;
4. Repeat until $M = \emptyset$;
   if $\rho(J_R^T) + \max(\rho(J_C), nnz(r)) < (\rho(J_C) + \max(\rho(J_R^T), nnz(c)))$;
      $J_C = J_C \bigcup (r \bigcap C)$;
      $R = R - \{r\}$
   else
      $J_R = J_R \bigcup (c \bigcap R)$;
      $C = C - \{c\}$;
   end if
   $M = J(R, C)$.
   end repeat

Note that, on completion, J_R, J_C have been defined; the requisite permutation matrices are implicitly defined by the ordering chosen in **MNCO**.

2.4 ▪ Sparse Jacobian-Matrix Computation

In this section, we illustrate how to evaluate a sparse Jacobian matrix using ADMAT. We give detailed descriptions of two popular sparsity functions: getjpi and evalj.

Description of getjpi

Function getjpi computes the sparsity information for the efficient computation of a sparse Jacobian matrix. It is invoked as follows:

```
[JPI, SPJ] = getjpi(fun, n, m, Extra, method, SPJ)
```

Input arguments

fun is the function to be differentiated; a string variable representing the function name.
n is the number of columns of the Jacobian matrix; a scalar.

m is the number of rows of the Jacobian matrix; by default, m = n; a scalar.
Extra stores parameters required in function fun apart from the independent variable; a MATLAB cell structure.
method sets techniques used to get the coloring information:

- method = 'd': direct bi-coloring (the default).
- method = 's': substitution bi-coloring.
- method = 'c': one-sided column method.
- method = 'r': one-sided row method.
- method = 'f': sparse finite-difference.

Detailed background on the various methods is given in [39, 95].

SPJ is the user-specified sparsity pattern of the Jacobian in MATLAB sparse format.

Output arguments

JPI includes the sparsity pattern, coloring information, and other information required for efficient Jacobian computation.

SPJ is the sparsity pattern of the Jacobian, represented in the MATLAB sparse format.

Note that ADMAT computes the sparsity pattern, SPJ, of the Jacobian first before calculating JPI. This is an expensive operation. If the user already knows the sparsity pattern SPJ, then it can be passed to the function getjpi as an input argument, so that ADMAT will calculate JPI based on the user-specified sparsity pattern. There is no need to recalculate the sparsity pattern, SPJ; it is inefficient to do so.

Description of evalj

Function evalj computes the sparse Jacobian matrix based on the sparsity pattern and coloring information obtained from getjpi. It is invoked as follows:

```
[F, J] = evalj(fun, x, Extra, m, JPI, verb, stepsize)
```

Input arguments

fun is the function to be differentiated; a string representing the function name.
x is the vector of independent variables; a vector.
Extra contains parameters required in function fun; a MATLAB cell structure.
m is the row dimension of the vector mapping, i.e., $F : R^n \rightarrow R^m$; a scalar.
JPI is the sparsity pattern and coloring information recorded for the current sparse Jacobian Jacobian.
verb holds flags for display level.

- verb = 0: no display (the default).
- verb = 1: display the number of groups used.
- verb $\geq$ 2: display in graph upon termination.

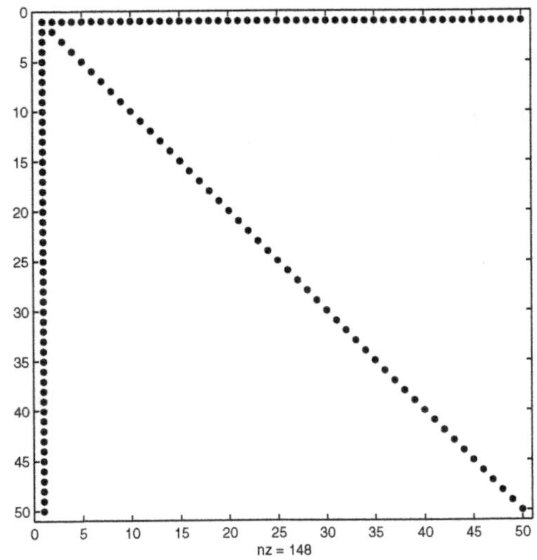

Figure 2.9. *Jacobian sparsity pattern of* arrowfun.

stepsize is the step size for finite-differencing and is needed only when JPI is computed by the finite-difference method, f. By default, stepsize = 1e-5.

Output arguments

F is the function value at point x; a vector.

J is the Jacobian matrix at point x; a sparse matrix.

Example 2.3. Compute the Jacobian matrix of an arrowhead function.

See DemoSprJac.m

Let $y = F(x)$, $F : R^n \to R^n$, where

$$y(1) = 2x(1)^2 + \sum_{i=1}^{n} x(i)^2,$$
$$y(i) = x(i)^2 + x(1)^2, \quad i = 2:n.$$

The following is the arrowhead function in MATLAB:

```
function y= arrowfun(x,Extra)
   y = x.*x;
   y(1) = y(1)+x'*x;
   y = y + x(1)*x(1);
```

The corresponding Jacobian matrix has an arrowhead sparsity structure, as shown in Figure 2.9, for $n = 50$ with 148 nonzeros. The procedure to evaluate the Jacobian J at an "all ones" vector for $n = 5$ is as follows.

1. Set problem size.
```
>> n = 5;
```

2. Initialize x.
```
>> x = ones(n,1)
x =
    1
    1
    1
    1
    1
```

3. Compute JPI corresponding to function arrowfun.
```
>> JPI = getjpi('arrowfun', n);
```

4. Compute the function value and the Jacobian (in MATLAB sparse format) based on the computed JPI. Set the input argument, Extra, to "[ ]" since it is empty.
```
>> [F, J] = evalj('arrowfun', x, [], n, JPI)

F =
    7
    2
    2
    2
    2
J =
    (1,1)  6
    (2,1)  2
    (3,1)  2
    (4,1)  2
    (5,1)  2
    (1,2)  2
    (2,2)  2
    (1,3)  2
    (3,3)  2
    (1,4)  2
    (4,4)  2
    (1,5)  2
    (5,5)  2
```

■

The function getjpi determines the sparsity pattern of the Jacobian of arrowfun. It needs to be executed only once for a given function. In other words, once the sparsity pattern is determined by getjpi, ADMAT calculates the Jacobian at a given point based on the pattern.

2.5 ▪ Sparse Hessian Computation

The process for determining sparse Hessian matrices is similar to the process for sparse Jacobians in ADMAT.

Description of gethpi

Function gethpi computes the sparsity pattern of a Hessian matrix and corresponding coloring information. It is invoked as follows:

[HPI, SPH] = gethpi(fun, n, Extra, method, SPH)

Input arguments

fun is the function to be differentiated; a string representing the function name.

n is the order of the Hessian; a scalar.

Extra stores parameters required in function fun (apart from the independent variable); a MATLAB cell structure.

method sets techniques used to get the coloring information.

- method = 'i-a': the default, ignore the symmetry; computing exactly using AD.
- method = 'd-a': direct method, using AD.
- method = 's-a': substitution method, using AD.
- method = 'i-f': ignore the symmetry and use finite differences (FD).
- method = 'd-f': direct method with FD.
- method = 's-f': substitution method with FD.

Note that details for the different methods are given in [95, 39].

SPH is the user-specified sparsity pattern of Hessian in MATLAB sparse format.

Output arguments

HPI includes the sparsity pattern, coloring information, and other information required for efficient Hessian computation.

SPH is sparsity pattern of Hessian, represented in the MATLAB sparse format.

Similar to getjpi, users can pass SPH to function gethpi when the sparsity pattern of Hessian is already known. The computation of HPI will be based on the userspecified sparsity pattern.

Description of evalh

[v, grad, H] = evalh(fun, x, Extra, HPI, verb)

Input arguments

fun is function to be differentiated; a string representing the function name.

x is the independent variable; a vector.

Extra stores parameters required in function fun.

HPI is the sparsity pattern and coloring information of Hessian.

verb is flag for display level.

- verb = 0: no display (the default).
- verb = 1: display the number of groups used.
- verb $\geq$ 2: display in graph upon termination.

Output arguments

v is the function value at point x; a scalar.

grad is the gradient at point x; a vector.

H is Hessian at point x; a sparse matrix.

Example 2.4. Compute the Hessian of the Brown function at point $x^T = [1, 1, 1, 1, 1]$.

See DemoSprHess.m

1. Set the problem size.

```
>> n = 5
```

2. Set the independent variable.

```
>> x = ones(n,1)
x =
    1
    1
    1
    1
    1
```

3. Compute the relevant sparsity information encapsulated in HPI.

```
>> HPI = gethpi('brown', n);
```

4. Evaluate the function value, gradient, and Hessian at x. We set the input argument Extra to "[ ]" since it is empty.

```
>> [v, grad, H] = evalh('brown', x, [], HPI)
v =
     8
grad =
     4     8     8     8     4
```

```
H =
   (1,1)  12
   (2,1)   8
   (1,2)   8
   (2,2)  24
   (3,2)   8
   (2,3)   8
   (3,3)  24
   (4,3)   8
   (3,4)   8
   (4,4)  24
   (5,4)   8
   (4,5)   8
   (5,5)  12
```

■

Similar to the sparse Jacobian situation, function **gethpi** encapsulates the sparsity structure and relevant coloring information for efficient calculation of the sparse Hessian H. Function **gethpi** needs to be executed only once for a given function. In other words, once the sparsity pattern is determined by **gethpi**, ADMAT calculates the Hessian at a given point based on the pattern.

Chapter 3

Using ADMAT with the MATLAB Optimization Toolbox

The MATLAB optimization toolbox includes solvers for many nonlinear problems, such as multidimensional nonlinear minimization, nonlinear least squares with upper and lower bounds, nonlinear system of equations, and so on. Typically, these solvers use derivative information such as gradients, Jacobians, and Hessians. In this chapter, we illustrate how to conveniently use ADMAT to accurately and automatically compute derivative information for use with the MATLAB Optimization Toolbox.

3.1 ▪ Nonlinear Equations Solver fsolve

MATLAB provides a function **fsolve** to solve nonlinear equations

$$F(x) = 0,$$

where $F : R^n \rightarrow R^n$. By default, this solver uses the finite difference method to estimate Jacobian matrices, but a user-defined Jacobian, or Jacobian information (e.g., when using the Jacobian-vector multiplication), can be specified as an alternative to finite differencing. Users just need to set up a flag in input argument **options** without changing any original code. The following example shows how to solve nonlinear equations with ADMAT computing Jacobian matrices. For the details of input and output arguments of the MATLAB solver **fsolve**, refer to MATLAB help documentation.

Example 3.1. Solving nonlinear equations using ADMAT to compute the Jacobian matrix.

See DemoFSolve.m

Solve the nonlinear equations,

$$F(x) = 0,$$

where $F(x)$ is the arrow function defined in Section 3.4.

1. Set problem size.

```
>> n = 5;
```

2. Initialize the random seed.

```
>> rand('seed' ,0);
```

3. Initialize x.

```
>> x0 = rand(n,1)
x0 =
   0.2190
   0.0470
   0.6789
   0.6793
   0.9347
```

4. Get the default value of **options** of MATLAB **fsolve** solver.

```
>> options = optimset('fsolve');
```

5. Turn on the Jacobian flag in input argument **options**. This means that the user will provide the method to compute Jacobians (in this case, the use of ADMAT).

```
>> options = optimset(options, 'Jacobian', 'on');
```

6. Set the function to be differentiated by ADMAT. The function call **feval** is overloaded by the one defined in ADMAT, which returns the function value and Jacobian on each **feval** call.

```
>> myfun = ADfun('arrowfun', n);
```

7. Call **fsolve** to solve the nonlinear least squares problem using ADMAT to compute derivatives.

```
>> [x, RNORM] = fsolve(myfun, x0, options)
x =
   0.0009
   0.0002
   0.0027
   0.0027
   0.0037

RNORM =
1.0e-004*
   0.2963
   0.0077
   0.0776
   0.0777
   0.1406
```

■

3.2 ▪ Nonlinear Least Squares Solver lsqnonlin

MATLAB provides a nonlinear least squares solver, **lsqnonlin**, for solving nonlinear least squares problems,

$$\min\left\{\|F(x)\|_2^2 \quad \text{s.t.} \quad l \leq x \leq u\right\},$$

where $F(x)$ maps R^m to R^n. By default this solver employs the finite-difference method to estimate gradients and Hessians (unless users specify analternative). ADMAT, for example, can be specified as an alternative to finite differences. Users just need to set up a flag in input argument **options** without changing any original codes. The following examples illustrate how to solve nonlinear least squares with ADMAT used to compute derivatives. For the details of input and output arguments of the MATLAB solver **lsqnonlin**, refer to MATLAB help documentation.

Example 3.2. Solving a nonlinear least squares problem using ADMAT to compute the Jacobian matrix.

See DemoLSq.m

Solve the nonlinear least squares problem,

$$\min\left\{\|F(x)\|_2^2 \quad \text{s.t.} \quad l \le x \le u\right\},$$

where $F(x)$ is the Broyden function defined in Section 2.3.

1. Set problem size.

```
>> n = 5;
```

2. Initialize the random seed.

```
>> rand('seed' ,0);
```

3. Initialize x.

```
>> x0 = rand(n,1)
x0 =
   0.2190
   0.0470
   0.6789
   0.6793
   0.9347
```

4. Set the lower bound for x.

```
>> l = -ones(n,1);
```

5. Set the upper bound for x.

```
>> u = ones(n,1);
```

6. Get the default value of options of MATLAB lsqnonlin solver.

```
>> options = optimset('lsqnonlin');
```

7. Turn on the Jacobian flag in input argument options. This means that the user will provide the method to compute Jacobians (in this case, the use of ADMAT).

```
>> options = optimset(options, 'Jacobian', 'on');
```

8. Set the function to be differentiated by ADMAT. The function call feval is overloaded by the one defined in ADMAT, which returns the function value and Jacobian on each feval call (see Section 2.3).

```
>> myfun = ADfun('broyden', n);
```

9. Call lsqnonlin to solve the nonlinear least squares problem using ADMAT to compute derivatives.

```
>> [x, RNORM] = lsqnonlin(myfun, x0, l,u, options)
x =
   -0.4107
   -0.2806
    0.2254
    1.0000
    1.0000
RNORM =
    1.0734
```

■

3.3 ▪ Multidimensional Nonlinear Minimization Solvers fmincon and fminunc

Consider a multidimensional constrained nonlinear minimization problem,

$$\begin{aligned} &\min f(x) \\ \text{s.t.}\quad & Ax \leq b, \quad Aeq * x = beq \quad \text{(linear constraints)}, \\ & c(x) \leq 0, \quad ceq(x) = 0 \quad \text{(nonlinear constraints)}, \\ & l \leq x \leq u. \end{aligned}$$

The MATLAB solver for this problem is fmincon.

The unconstrained minimization problem is simply

$$\min f(x),$$

where $f(x)$ maps R^n to scalars. This unconstrained problem can be solved by fminunc. ADMAT can be used to compute the gradient or both the gradient and the Hessian matrix.

Example 3.3. Solve unconstrained nonlinear minimization problems using ADMAT.

See DemoFminunc.m

Solve the nonlinear minimization problem,

$$\min \{brown(x)\},$$

where $brown(x)$ is the Brown function defined in Section 3.1. In this example, we will solve the problem twice. In the first solution, ADMAT is used to compute gradients only, and Hessians are estimated by the default finite-difference method. In the second solution, ADMAT is used to compute both gradients and Hessians.

1. Set problem size.

```
>> n = 5;
```

2. Initialize the random seed.

```
>> rand('seed', 0);
```

3. Initialize x.

```
>> x0 = rand(n,1)
x0 =
    0.2190
    0.0470
    0.6780
    0.6793
    0.9347
```

4. Set the function to be differentiated by ADMAT. The function call feval is overloaded by the one defined in ADMAT. It can return the function value, gradient, and Hessian in each feval call (see Section 3.1 for details).

```
>> myfun = ADfun('brown',1);
```

5. Solve the problem using ADMAT to determine gradients (but not Hessians).

   (a) Turn on the gradient flag in input argument options (but not the Hessian flag). Thus, the solver will use ADMAT to compute gradients but will estimate Hessians by the finite-difference method.

   ```
   >> options = optimoptions(@fminunc, 'GradObj', 'on', 'Hessian', 'off',
   'Algorithm', 'trust-region');
   ```

   (b) Call the MATLAB constrained nonlinear minimization solver fminunc with ADMAT used to determine gradients only.

   ```
   >> [x,FVAL] = fminunc(myfun,x0,options)
   x =
     1.0e-007 *
       0.0059
       0.2814
      -0.8286
       0.2189
       0.2832
   FVAL =
     1.7076e-14
   ```

6. Solve the problem using ADMAT to compute both gradients and Hessians.

   (a) Turn on both gradient and Hessian flags in input argument options. Thus, the solver will use the user-specified method (ADMAT) to compute both gradients and Hessians.

   ```
   >> options = optimoptions(@fminunc, 'GradObj', 'on', 'Hessian', 'on',
   'Algorithm', 'trust-region');
   ```

(b) Call the MATLAB constrained nonlinear minimization solver **fmincon** using ADMAT to compute derivatives.

```
>> [x,FVAL] = fminunc(myfun,x0,options)
x =
  1.0e-007 *
     0.0059
     0.2814
    -0.8286
     0.2189
     0.2832
FVAL =
  1.7076e-14
```

■

Example 3.4. Solve the constrained nonlinear minimization problems using ADMAT.

See DemoFmincon.m

Solve the nonlinear minimization problem,

$$\min brown(x), \quad l \leq x \leq u,$$

where $brown(x)$ is the Brown function defined in (4.38).

1. Set problem size.

```
>> n = 5;
```

2. Initialize the random seed.

```
>> rand('seed', 0);
```

3. Initialize x.

```
>> x0 = rand(n,1)
x0 =
    0.2190
    0.0470
    0.6789
    0.6793
    0.9347
```

4. Set the lower bound for x.

```
>> l = -ones(n,1);
```

5. Set the upper bound for x.

```
>> u = ones(n,1);
```

6. Set the function to be differentiated by ADMAT.

   ```
   >> myfun = ADfun('brown',1);
   ```

7. Set up **options** so that both gradients and Hessians are computed by ADMAT.

   ```
   >> options = optimoptions(@fmincon, 'GradObj', 'on', 'Hessian', 'on',
   'Algorithm', 'trust-region-reflective');
   ```

8. Call MATLAB constrained nonlinear minimization solver **fmincon** with ADMAT to compute derivatives.

   ```
   >> [x,FVAL] = fmincon(myfun,x0,[],[],[],[],l,u,[], options)
   x =
     1.0e-007 *
      -0.0001
       0.0000
      -0.1547
      -0.1860
       0.0869
   FVAL =
     1.2461e-015
   ```

In summary, ADMAT can be conveniently linked to MATLAB nonlinear least squares solver lsqnonlin and nonlinear minimization solvers fminunc and fmincon in two steps:

1. Set up the Jacobian, gradient, and Hessian flags in the input argument options.

2. Set the function to be differentiated by ADMAT using the ADfun function call to overload feval.

■

3.4 ▪ Additional Applications of AD

In this section we illustrate two additional applications of ADMAT. First, we show how to trigger the quasi-Newton computation in the MATLAB unconstrained nonlinear minimization solver fminunc with ADMAT used to determine gradients. Second, we present a sensitivity problem.

3.4.1 ▪ Quasi-Newton Computation

The MATLAB multidimensional unconstrained nonlinear minimization solver fminunc, uses the quasi-Newton approach when the user chooses the medium-scale option (to classify the size of the problem being solved). This method updates the inverse of Hessian directly by the Broyden–Fletcher–Goldfarb–Shanno (BFGS) formula and, by default, estimates gradients by finite differencing. However, it allows users to specify their own gradient computation method to replace the finite-difference method. The following example shows how to trigger the quasi-Newton computation in fminunc with ADMAT used to determine gradients.

Example 3.5. Find a minimizer of the Brown function using the quasi-Newton approach in fminunc (with gradients determined by ADMAT).

See DemoQNFminunc.m

1. Set the problem size.

```
>> n = 5;
```

2. Initialize the random seed.

```
>> rand('seed',0);
```

3. Initialize x.

```
>> x0 = rand(n,1)
x0=
    0.2190
    0.0470
    0.6789
    0.6793
    0.9347
```

4. Set the function to be differentiated by ADMAT.

```
>> myfun = ADfun('brown',1);
```

Note: The second input argument in ADfun, 1, is a flag indicating a scalar mapping, $f: R^n \rightarrow R^1$; more generally, the second argument is set to m for a vector-valued function, $F: R^n \rightarrow R^m$.

5. Get the default value of argument options.

```
>> options = optimset('fmincon');
```

6. Turn on the gradient flag of input argument options. Thus, the solver uses a user-specified method to compute gradients (i.e., ADMAT).

```
>> options = optimset(options, 'GradObj', 'on');
```

7. Set the flag of LargeScale to off, so that the quasi-Newton method will be used.

```
>> options = optimset(options, 'LargeScale', 'off');
```

8. Solve the constrained nonlinear minimization by the quasi-Newton method in fmincon with ADMAT used to compute gradients.

```
>> [x,FVAL] = fminunc(myfun,x0, options)
x =
   1.0e-006 *
```

```
        —0.2034
        —0.5461
        —0.1301
        —0.3328
         0.1859
FVAL =
    9.2779e-013
```

We remark that this combination—AD with BFGS—can be an efficient way to solve a local minimization problem. The computing cost at each iteration is proportional to $\omega(f)+O(n^2)$. This contrasts with the typical BFGS approach (i.e., no use of AD), when each iteration costs $n \cdot \omega(f)+O(n^2)$. ∎

3.4.2 ▪ A Sensitivity Problem

This example is concerned with sensitivity with respect to problem parameters at the solution point. Consider a nonlinear scalar-valued function $f(x,\mu)$, where x is the independent variable and μ is a vector of problem parameters. This example illustrates that ADMAT can be used both to compute derivatives with respect to x in order to minimize $f(x,\mu)$ with respect to x (for a fixed value of μ) and to analyze the sensitivity of $f(x_{opt},\mu)$ with respect to μ at the solution point.

Example 3.6. Analyze the sensitivity of brownv(x, V) function with respect to μ at the optimal point x_{opt} with $V=0.5$.

See DemoSens.m

The function brownv(x, V) is similar to the Brown function given in (4.38). Its definition is as follows:

```
function f = brownv(x,V)
% length of input x
n=length(x);
% if any input is a 'deriv' class object, set n to 'deriv' class as

if isa(x, 'deriv') || isa(V, 'deriv')
    n = deriv(n);
end
y=zeros(n,1);
i=1:(n-1);
y(i)=(x(i).^2).^(V*x(i+1).^2+1)+((x(i+1)+0.25).^2).^(x(i).^2+1) + ...
    (x(i)+0.2*V).^2;
f = sum(y);
tmp = V'*x;
f = f - .5*tmp'*tmp;
```

1. Set the problem size.

   `>> n = 5;`

2. Set the initial value of V to 0.5. We first solve the minimization problem, $\min \text{brownv}(x, V)$, at $V = 0.5$, then analyze the sensitivity of $\text{brownv}(x_{opt}, V)$ with respect to V.

```
>> V = 0.5
```

3. Initialize the random seed.

```
>> rand('seed',0);
```

4. Set the initial value of x.

```
>> x0 = 0.1*rand(n,1)
x0 =
    0.2190
    0.0470
    0.6789
    0.6793
    0.9347
```

5. Solve the unconstrained minimization problem, $\min \text{brownv}(x, V)$, using the MATLAB optimization solver **fminunc** and with ADMAT as a derivative computation method to compute gradients and Hessians (see Section 6.2 for details).

```
>> options = optimset('fminunc');
>> myfun = ADfun('brownv', 1);
>> options = optimset(options, 'GradObj', 'on');
>> options = optimset(options, 'Hessian', 'on');
>> [ x1, FVAL1 ] = fminunc(myfun, x0, options, V);
```

6. Set the parameter V to **deriv** class.

```
>> V = deriv(V,1);
```

7. Compute the function value of $\text{brownv}(x, V)$ at the optimal point $x1$ with **deriv** class objective V.

```
>> f = brownv(x1, V)
val =
     0.0795
deriv =
     -0.0911
```

8. Get the sensitivity of V at optimal point x_{opt}.

```
>> sen = getydot(f)
sen =
     -0.0911
```

■

In summary, analyzing the sensitivity $f(x,\mu)$ with respect to μ at $x = x_{opt}$ requires two steps:

- Solve the minimization problem of $f(x,\mu)$ with respect to x at a fixed value μ.
- Differentiate the function $f(x,\mu)$ with respect to μ at the optimal point x_{opt} to get the sensitivity of $f(x_{opt},\mu)$ with respect to μ.

Chapter 4

Structure

Many practical optimization problems, as well as nonlinear system problems, exhibit structure. This structure can show up in the form of sparsity, discussed in the previous chapter, but not always. Indeed, many compute-intensive problems exhibit limited sparsity in their Jacobian and/or Hessian matrices. However, in many of these cases the objective function exhibits structure that can be used to great advantage when computing gradients, derivative matrices, or Newton steps. *This structure often hides a deeper sparsity, and, consequently, sparse derivative techniques can often be used to great efficiency advantage on these dense (but structured) problems.*

We begin with an example, from [44], to illustrate this widely applicable idea. Suppose that the evaluation of $z = F(x)$ is a structured computation given by $F^E(x, y_1, y_2)$, defined by the following three macro-steps:

$$\left.\begin{array}{llll} (1) & \text{Solve for } y_1 & : & y_1 - \tilde{F}(x) = 0 \\ (2) & \text{Solve for } y_2 & : & Ay_2 - y_1 = 0 \\ (3) & \text{Solve for } z & : & z - \bar{F}(y_2) = 0 \end{array}\right\}, \tag{4.1}$$

where $\tilde{F}$, $\bar{F}$ are different vector-valued functions of x and y_2, respectively, and $A = A(x)$ is an invertible matrix with smooth differentiable entries. The Newton step at point x is $s_N = -J^{-1}F$, with J, F evaluated at the current point x. It is straightforward to see that

$$J = \bar{J}A^{-1}[\tilde{J} - A_x y_2]. \tag{4.2}$$

We can deduce from (4.2) that regardless of any sparsity presented in the matrices $\bar{J}$, A, $\tilde{J}$, $A_x y_2$, the Jacobian matrix J will be dense (almost surely). This dense structure is due to the application of A^{-1}. Since the Jacobian J is dense, it is not at all apparent that sparse AD techniques can be used, and the work required to compute J by automatic differentiation (or indeed finite differencing) would therefore be proportional to $n \cdot \omega(F)$, where $\omega(F)$ is the work required to evaluate $F(x)$. Note that $\omega(F) = \omega(F^E)$.

However, it is possible to use sparse AD techniques on this problem, and it is well worthwhile to do this. The key is to realize that the Jacobian of $F^E(x, y_1, y_2)$, where F^E is defined by (4.1), is sparse and that sparse AD techniques can be applied to $F^E(x, y_1, y_2)$ to obtain the sparse Jacobian

$$J^E = \begin{bmatrix} -\tilde{J} & I & 0 \\ A_x y_2 & -I & A \\ 0 & 0 & \bar{J} \end{bmatrix}. \tag{4.3}$$

The cost to obtain J^E is therefore proportional to $\chi(J^E) \cdot \omega(F)$, where $\chi(J^E)$ is a "chromatic number" of a graph reflecting the sparsity of J^E, usually $\chi(J^E) \ll n$. Of course it is reasonable to ask, "What do I do with matrix J^E?" There are two answers.

First, it is possible to compute J, given J^E, by doing a bit of linear algebra (i.e., a Schur-complement computation). Specifically, $J = \bar{J}A^{-1}[\tilde{J} - A_x y_2]$, and, as illustrated in [44], the computation of J can be relatively insignificant compared to the determination of J. Second, also as indicated in [44], if it is the Newton step that is required, then explicit determination of J given J^E may not be the best way to go. Specifically, it may be preferable to directly solve

$$J^E \begin{pmatrix} \delta x \\ \delta y_1 \\ \delta y_2 \end{pmatrix} = \begin{pmatrix} 0 \\ 0 \\ -F(x) \end{pmatrix}, \tag{4.4}$$

using a sparse matrix solve technique, where it turns out that δx is the Newton step (i.e., $J \cdot \delta x = -F(x)$). Refer to Chapter 5 for details.

4.1 ▪ The General Situation for $F : \mathscr{R}^n \to \mathscr{R}^m$

How common is the situation illustrated by example (4.1)? It turns out that (4.1) has a straightforward generalization, and this generalization is very common. The point of the generalization is that it allows for the application of sparse matrix AD techniques to efficiently get Jacobian matrices and Newton steps on dense problems. The efficiency gains can be very significant.

Suppose that at iterate x, the computation $z = F(x)$ can be written as follows:[4]

$$\left.\begin{array}{lcl} \text{Solve for } y_1 & : & F_1^E(x, y_1) = 0 \\ \text{Solve for } y_2 & : & F_2^E(x, y_1, y_2) = 0 \\ \vdots & & \vdots \\ \text{Solve for } y_p & : & F_p^E(x, y_1, y_2, \ldots, y_p) = 0 \\ \text{Solve for output } z & : & z - F_{p+1}^E(x, y_1, y_2, \ldots, y_p) = 0 \end{array}\right\}. \tag{4.5}$$

Then, assuming that the Jacobian of each successive function $F_i^E(x, y_1, \ldots, y_i)$, $i = 1, \ldots, p$, is sparse, the sparse AD techniques can be applied to F_i^E for each successive i, in turn.

[4]Note that the dimension of y_i can vary. In addition, a practical working assumption is that p is modest (i.e., $p \ll \omega(F)$), and the component functions F_i^E $(i = 1 : p)$ are accessible.

Hence, the Jacobian

$$J^E = \left(\begin{array}{c|ccccc} \frac{\partial F_1}{\partial x} & \frac{\partial F_1}{\partial y_1} & & & & \\ \frac{\partial F_2}{\partial x} & \frac{\partial F_2}{\partial y_1} & \frac{\partial F_2}{\partial y_2} & & & \\ \vdots & \vdots & \vdots & \ddots & & \\ \frac{\partial F_p}{\partial x} & \frac{\partial F_p}{\partial y_1} & \cdots & \cdots & \frac{\partial F_p}{\partial y_p} \\ \hline \frac{\partial F_{p+1}}{\partial x} & \frac{\partial F_{p+1}}{\partial y_1} & \cdots & \cdots & \frac{\partial F_{p+1}}{\partial y_p} \end{array}\right) \tag{4.6}$$

is determined by applying sparse AD techniques to obtain each successive block row in turn. The work required to obtain J^E in this fashion is thus proportional to

$$\sum \chi(J_i^E) \cdot \omega(F_i^E) \leq \chi(J^E) \cdot \omega(F^E) = \chi(J^E) \cdot \omega(F), \tag{4.7}$$

where J_i^E is the Jacobian of $F_i^E(x, y_1, \ldots, y_p)$, $\chi(J_i^E)$ is the "chromatic number" of a graph reflecting the sparsity of J_i^E [42], and $\chi(J^E)$ is the "chromatic number" of the graph reflecting the sparsity of J^E. Therefore, if $\chi(J^E) \ll n$, then the work required to compute J^E can be significantly less than the work required to compute J by applying AD to F in a direct way (the cost of the direct approach is $n \cdot \omega(F)$).

If we write (4.6) as

$$J^E = \left(\begin{array}{c|c} \mathrm{A} & \mathrm{B} \\ \hline \mathrm{C} & \mathrm{D} \end{array}\right),$$

then the Jacobian of F is given by the Schur-complement computation,

$$J = C - DB^{-1}A. \tag{4.8}$$

Typically, J^E exhibits block sparsity; in addition, the nonzero blocks of J^E in (4.6) are often sparse themselves.

We note again that if it is the Newton step required, then it can be practical to solve the large sparse system (4.4) for the Newton step s_N. See Chapter 5 for more details on this direct Newton approach.

Example 4.1. A composite function is a highly recursive function, $F : \mathcal{R}^n \to \mathcal{R}^m$:

$$F(x) = \bar{F}(T_p(T_{p-1}(\cdots(T_1(x))\cdots))), \tag{4.9}$$

where functions $\bar{F}$ and T_i $(i = 1, \ldots, p)$ are vector maps. A dynamical system is a special case of (4.9) where each transformation is identical, i.e., $T_i = T (i = 1, \ldots, p)$.

The Jacobian of F is a matrix product

$$J = \bar{J} \cdot J_p \cdot J_{p-1} \cdot \cdots \cdot J_1, \tag{4.10}$$

where J_i is the Jacobian of T_i evaluated at $T_{i-1}(T_{i-2}(\cdots(T_1(x))\cdots))$ and $\bar{J}$ is the Jacobian of $\bar{F}$ evaluated at $T_p(T_{p-1}(\cdots(T_1(x))\cdots))$. The general structured form (4.5) can be used, naturally, to express this recursive computation in (4.7):

$$\begin{array}{l} \text{for } i = 1, \ldots, p \\ \qquad \text{Solve for } y_i : \ y_i - T_i(y_{i-1}) = 0, \\ \text{Solve for } z : \ z - \bar{F}(y_p) = 0, \end{array} \tag{4.11}$$

where $y_0 = x$.

In this case, the extended Jacobian matrix J^E simplifies to

$$J^E = \begin{pmatrix} -J_1 & I & & & \\ & -J_2 & I & & \\ & & \ddots & \ddots & \\ & & & -J_p & I \\ & & & & \bar{J} \end{pmatrix}. \tag{4.12}$$

■

This example is a tightly coupled recursive computation. To contrast, consider the following class of partially separable problems discussed in Section 6.5.2.

Example 4.2. In contrast to the composite function and dynamical systems case, a generalized partially separable function represents a function with little recursion and a plentitude of independent computations. Generalized partially separable functions are also very common; moreover, this class is also covered by our notion of structure.

A square generalized partially separable (GPS) function is a vector mapping, $F : \mathcal{R}^n \rightarrow \mathcal{R}^n$, defined by

$$\left.\begin{array}{l} y_i = T_i(x), \quad i = 1, \ldots, p \\ F(x) = \bar{F}(y_1, \ldots, y_p) \end{array}\right\}. \tag{4.13}$$

It is trivial to put this computation into the general form defined by (4.5):

$$\begin{array}{l} \text{for } i = 1, \ldots, p \\ \qquad \text{Solve for } y_i : \ y_i - T_i(x) = 0 \\ \text{Solve for } z : \ z - \bar{F}(y_1, \ldots, y_p) = 0. \end{array} \tag{4.14}$$

In this case the auxiliary Jacobian matrix structure (4.6) reduces to

$$J^E = \left(\begin{array}{c|cccc} -J_1 & I & & & \\ -J_2 & & I & & \\ \vdots & & & \ddots & \\ -J_p & & & & I \\ \hline 0 & \bar{J}_1 & \bar{J}_2 & \cdots & \bar{J}_p \end{array}\right), \tag{4.15}$$

where $\bar{J}$ is divided into different components $\bar{J}_i$ corresponding to variable vector y_i and J_i, the Jacobian of T_i, $i = 1, \ldots, p$. Note that J^E is highly structured and sparse. In fact, in most applications there will be additional sparsity in both J_i and $\bar{J}_i$. For example, $\bar{F}$ often represents the summation operation, in which case $\bar{J}_i$ is the identity matrix; T_i is often an

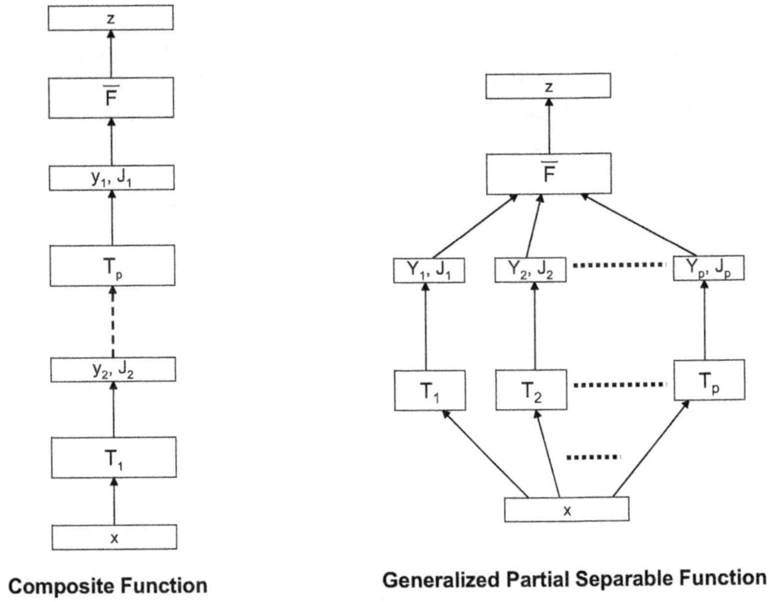

Figure 4.1. *Block diagrams of composite function and generalized partial separable function. (See figures* 3 *and* 4 *in* [42]*).*

active function of just a small subset of the components of x. Hence, the work to obtain J^E is $\chi \cdot \omega(F)$, where χ is often quite small (e.g., $\chi < 10$). ∎

These two examples have contrasting structure, illustrated in Figure 4.1. However, both are captured by the general structure given in (4.5).

4.2 ▪ The Gradient Revisited

When the nonlinear function $F : \mathcal{R}^m \to \mathcal{R}^m$ is available in a structured form (4.5), it is usually advantageous to apply AD, "slice by slice," obtaining each block row of the extended Jacobian matrix J^E in order. That is, at step i, compute the Jacobian of function $F_i^E(x, y_1, \ldots, y_{i-1})$. This approach is efficient because a combination of forward and reverse modes can be used to effectively determine the (usually) sparse Jacobian J_i^E of F_i^E while working only with the computational tape corresponding to function F_i^E. We note that all preceding tape segments corresponding to F_i^E can be discarded since they are no longer required; the sparse matrices J_i^E are of course saved, as they are computed. The true Jacobian $J \in \mathcal{R}^{m \times n}$ can be calculated from J^E using the Schur-complement formulation (4.8). Alternatively, it is sometimes more efficient to use the extended Jacobian directly to simulate the action of J (e.g., Chapter 5).

These structural ideas apply to the case $m = 1$, which covers the gradient situation. That is, if $f : \mathcal{R}^n \to \mathcal{R}^1$, then the Jacobian of f is usually denoted as the gradient (transpose), and $(\nabla f)^T \in \mathcal{R}^{1 \times n}$. Due to the significant space savings and subsequent localized memory references, this structural approach, which includes computing and saving the sparse extended Jacobian matrix J^E, can produce realized efficiency gains over straight reverse-mode AD [32, 38, 39, 42]. In theory, an additional measure of the sparsity of J^E plays a role in the computing time of the gradient (versus straight reverse-mode AD). However, it was shown in [102] how to use the structured form without actually explicitly computing the extended Jacobian J^E. The result is a method with the same theoretical order of work as reverse-mode AD (i.e., not sparsity dependent), but it does not require as much memory as the reverse-mode AD. We discuss this approach next.

We begin by slightly narrowing the class of structured problems. Assume the scalar-valued function $z = f(x)$ is given in the following structured form:

$$\left.\begin{array}{l} \text{Solve for } y_1: \ F_1(x,y_1) \equiv \tilde{F}_1(x) - M_1 \cdot y_1 = 0 \\ \text{Solve for } y_2: \ F_2(x,y_1,y_2) \equiv \tilde{F}_2(x,y_1) - M_2 \cdot y_2 = 0 \\ \vdots \\ \text{Solve for } y_p: \ F_p(x,y_1,y_2,\ldots,y_p) \equiv \tilde{F}_p(x,y_1,y_2,\ldots,y_{p-1}) - M_p \cdot y_p = 0 \\ \text{Solve for } z: \ F_{p+1}(x,y_1,y_2,\ldots,y_{p+1}) \equiv \bar{f}(x,y_1,y_2,\ldots,y_p) - z = 0 \end{array}\right\}, \tag{4.16}$$

i.e., solve $F^E(x,y) = 0$, where each vector-valued function $\tilde{F}_i(x,y_1,\ldots,y_{i-1})$ is differentiable, as is the scalar-valued function $\bar{f}(x,y_1,\ldots,y_p)$. Each matrix M_i is nonsingular with the order of M_i equal to the dimension of vector y_i for $i = 1,\ldots,p$. Note that the intermediate vectors $y_1,\ldots,y_p$ are of varying lengths (in general). The corresponding extended Jacobian (which we will not explicitly compute in this approach) is

$$J^E = \begin{pmatrix} J_x^1 & -M_1 & & & \\ J_x^2 & J_{y_1}^2 & -M_2 & & \\ \vdots & \vdots & \vdots & \ddots & \\ J_x^p & J_{y_1}^p & \vdots & J_{y_{p-1}}^p & -M_p \\ \nabla\bar{f}_x^T & \nabla\bar{f}_{y_1}^T & \cdots & \cdots & \nabla\bar{f}_{y_p}^T \end{pmatrix}. \tag{4.17}$$

If we partition J^E as

$$J^E = \left(\begin{array}{c|cccc} J_x^1 & -M_1 & & & \\ J_x^2 & J_{y_1}^2 & -M_2 & & \\ \vdots & \vdots & \vdots & \ddots & \\ J_x^p & J_{y_1}^p & \vdots & J_{y_{p-1}}^p & -M_p \\ \hline \nabla\bar{f}_x^T & \nabla\bar{f}_{y_1}^T & \cdots & \cdots & \nabla\bar{f}_{y_p}^T \end{array}\right) = \left(\begin{array}{c|c} A & B \\ \hline \nabla\bar{f}_x^T & \nabla\bar{f}_y^T \end{array}\right),$$

then the gradient of f, with respect to x, satisfies

$$\nabla f^T = \nabla\bar{f}_x^T - \nabla\bar{f}_y^T B^{-1} A. \tag{4.18}$$

The key to the gradient computation idea given in [102] is to first compute and save vector y while evaluating $f(x)$ without doing any differentiation, then evaluate $(\nabla\bar{f}_x^T, \nabla\bar{f}_y^T)$ by applying reverse-mode AD to $\bar{f}(x,y)$, and finally complete the calculation of (4.31) by judiciously employing reverse-mode AD without ever explicitly computing the off-diagonal matrices $J_{y_i}^i$.

Specifically, consider the following approach, given in [104], to the computation of $v^T = \nabla \bar{f}_y^T B^{-1} A$. We can define $w^T = (w_1^T, \ldots, w_p^T)$ to satisfy $w^T = \nabla \bar{f}_y^T B^{-1}$, or

$$\begin{pmatrix} -M_1 & (J_{y_1}^2)^T & (J_{y_1}^3)^T & \cdots & (J_{y_1}^{p-1})^T & (J_{y_1}^p)^T \\ & -M_2 & (J_{y_2}^3)^T & \cdots & \vdots & (J_{y_2}^p)^T \\ & & \ddots & \cdots & \vdots & \vdots \\ & & & -M_{p-2} & (J_{y_{p-2}}^{p-1})^T & J_{y_{p-2}}^p \\ & & & & -M_{p-1} & J_{y_{p-1}}^p \\ & & & & & -M_p \end{pmatrix} \begin{pmatrix} w_1 \\ w_2 \\ w_3 \\ \vdots \\ w_{p-1} \\ w_p \end{pmatrix} = \begin{pmatrix} \nabla \bar{f}_{y_1} \\ \nabla \bar{f}_{y_2} \\ \nabla \bar{f}_{y_3} \\ \vdots \\ \nabla \bar{f}_{y_{p-1}} \\ \nabla \bar{f}_{y_p} \end{pmatrix} \tag{4.19}$$

and

$$v^T = \nabla \bar{f}^T B^{-1} A = w^T A = w_1^T J_x^1 + w_2^T J_x^2 + \cdots + w_{p-1}^T J_x^{p-1} + w_p^T J_x^p. \tag{4.20}$$

Thus, we can use the following algorithm to compute the structured gradient in (4.18).

ALGORITHM 4.1.
Algorithm for Structured Gradient

1. Follow (4.16) to evaluate the values of $y_1, y_2, \ldots, y_p$ only.
2. Evaluate $z = \bar{f}(x, y_1, \ldots, y_p)$ and apply reverse-mode AD to obtain $\nabla \bar{f}^T = (\nabla \bar{f}_x^T, \nabla \bar{f}_{y_1}^T, \ldots, \nabla \bar{f}_{y_p}^T)$.
3. Compute the gradient using (4.19):
   (a) Initialize $v_i = 0$, $i = 1 : p$, $\nabla f = \nabla \bar{f}_x^T$.
   (b) For $j = p, p-1, \ldots, 1$,
   - Solve $M_j w_j = \nabla \bar{f}_{y_j} - v_j$;
   - Evaluate $F_j(x, y_1, \ldots, y_{j-1})$ and apply reverse-mode AD with matrix w_j^T to get $w_j^T \cdot (J_x^j, J_{y_1}^j, \cdots, J_{y_{j-1}}^j)$;
   - Set $v_i^T = v_i^T + w_j^T \cdot J_{y_i}^j$ for $i = 1, \ldots, j-1$;
   - Update $\nabla f^T \leftarrow \nabla f^T + w_j^T J_x^j$.

The work in Algorithm 4.1 is proportional to $\omega(\bar{f}) + \sum_{i=1}^p \omega(F_i) = \omega(F)$. Space needs satisfy $\sigma \leq \max\{\omega(F_i),\ i = 1, \ldots, p\}$.

4.3 ▪ A Mixed Strategy for Computing Jacobian Matrices

The efficient structured gradient idea discussed above takes advantage of the structured extended Jacobian J^E without actually computing J^E. This method makes p calls to reverse-mode AD, each time applying reverse-mode AD with one of the component functions F_i and a pre-multiplying matrix W_i. Efficiency is gained because the row dimension of W_i is equal to unity in this gradient case. However, it is not hard to see that while $m = 1$ is the best case, this idea will gain advantage whenever m is less than the chromatic number (see

Section 4.1) of the graph corresponding to J_i^E. Recall that our structured form to evaluate $z = f(x)$ is

$$\left.\begin{array}{l}\text{Solve for } y_1: F_1(x,y_1) \equiv \tilde{F}_1(x) - M_1 \cdot y_1 = 0 \\ \text{Solve for } y_2: F_2(x,y_1,y_2) \equiv \tilde{F}_2(x,y_1) - M_2 \cdot y_2 = 0 \\ \vdots \\ \text{Solve for } y_p: F_p(x,y_1,y_2,\ldots,y_p) \equiv \tilde{F}_p(x,y_1,y_2,\ldots,y_{p-1}) - M_p \cdot y_p = 0 \\ \text{Solve for } z: F_{p+1}(x,y_1,y_2,\ldots,y_{p+1}) \equiv f(x,y_1,y_2,\ldots,y_p) - z = 0 \end{array}\right\}, \tag{4.21}$$

i.e., solve $F^E(x,y) = 0$, with corresponding extended Jacobian matrix

$$J^E = \left(\begin{array}{c|ccccc} J_x^1 & -M_1 & & & & \\ J_x^2 & J_{y_1}^2 & -M_2 & & & \\ \vdots & \vdots & \vdots & \ddots & & \\ J_x^p & J_{y_1}^p & \vdots & J_{y_{p-1}}^p & -M_p \\ \hline J_x^{p+1} & J_{y_1}^{p+1} & \cdots & \cdots & J_{y_p}^{p+1} \end{array}\right) \equiv \left(\begin{array}{c|c} A & B \\ \hline C & D \end{array}\right),$$

where

$$J_x^i = \frac{\partial F_i}{\partial x}\ (i = 1:p+1), \quad J_{y_j}^i = \frac{\partial F_i}{\partial y_j}\ (i = 2:p+1,\ j = 1:p), \quad M_i = \frac{\partial F_i}{\partial y_i},$$

and the Jacobian can be recovered from J^E, $J = C - DB^{-1}A$, as detailed below.

If $n < m$, then we compute J by first computing J^E block row by block row. That is, for $j = 1:p+1$, we first evaluate $F_j(x,y_1,\ldots,y_{j-1})$ apply sparse AD to $F_j(x,y_1,y_2,\ldots,y_{j-1})$ to get $(J_x^j, J_{y_1}^j, \ldots, J_{y_{j-1}}^j, M_j)$ and then solve $BV = A$. Recall that B is block lower triangular, and so only the diagonal blocks M_i need be factored. After V is determined, matrix multiplication and addition are needed to finally obtain $J = C - DV$.

If $m \leq n$, then the situation is more interesting because of the potential of using reverse-mode AD in a selective way. To begin, when $m \leq n$, it is advantageous to compute J by considering first the product DB^{-1} or the solution of

$$B^T W = D^T \tag{4.22}$$

for matrix $W = (W_1^T, \ldots, W_p^T)^T$, where W has m columns. Each submatrix W_i has the number of rows equal to the length of the corresponding vector y_i.

Similar to the case where $m = 1$, i.e., the gradient case, it may be advantageous to compute $W_j^T \cdot (J_x^j, J_{y_1}^j, \ldots, J_{y_{j-1}}^j)$ directly by reverse-mode AD in the process of solving (4.22). This is in contrast to using sparse AD to compute $(F_j)' = (J_x^j, J_{y_1}^j, \ldots, J_{y_{j-1}}^j)$ and then performing a matrix multiplication to get $W_j^T \cdot (J_x^j, J_{y_1}^j, \ldots, J_{y_{j-1}}^j)$. Let χ_j be the chromatic number of the graph of the sparse Jacobin $(F_j)^T = (J_x^j, J_{y_1}^j, \ldots, J_{y_{j-1}}^j)$. Then, if $\sum_{i=1}^p (\chi_i - m)^+ > p$, we first evaluate $F^E(x,y_1,\ldots,y_p)$ followed by the computation of F_{p+1} and $D^T = (F_{p+1})^T =$

$(\frac{\partial F_{p+1}}{\partial x}, \frac{\partial F_{p+1}}{\partial y_1}, \ldots, \frac{\partial F_{p+1}}{\partial y_p})$ by sparse AD. The following strategy computes W, efficiently mixing the two different ways (discussed above) of computing $W_j^T \cdot (J_x^j, J_{y_1}^j, \ldots, J_{y_{j-1}}^j)$:

For $j = p, p-1, \ldots, 1$
  Solve $M_j W_j = -\bar{J}_{y_i} - T_j$
  if $m \leq \chi_i$
    evaluate $F_j(x, y_1, \ldots, y_{j-1})$ and apply reverse-mode AD
    with matrix W_j^T to get $W_j \cdot (J_x^j, J_{y_1}^j, \ldots, J_{y_{j-1}}^j)$.
  else
    evaluate $F_j(x, y_1, \ldots, y_{j-1})$ and apply sparse AD to get
    $(J_x^j, J_{y_1}^j, \ldots, J_{y_{j-1}}^j)$ and multiply W_j to get $W_j \cdot (J_x^j, J_{y_1}^j, \ldots, J_{y_{j-1}}^j)$.
  Set $T_i = T_i + W_j \cdot J_{y_i}^j$ for $i = 1, \ldots, j-1$;
  Update $J \leftarrow J - W_j \cdot J_x^j$;

The space requirements for this approach still satisfy $\sigma \leq \max\{\omega(F_i),\ i = 1, \ldots, p\}$, but the work requirements satisfy

$$\omega(J) \sim \sum_{i=1}^{p} \min(\chi_i,\ m) \cdot \omega(F_i) + \sum_{i=1}^{p+1} \omega(F_i).$$

4.4 ▪ ADMAT Example on Structured Problems

In this section, we present some examples for implementing the structured ideas in the previous section via ADMAT. In each example, we use a function $g(x)$ as a basic function to create a "heavy" computing function for the gradient computation. The definition of $g(x)$ is

$$g(x) = \underbrace{f(f \cdots f}_{inner\text{-}p}\ (x)) \cdots), \tag{4.23}$$

where $f(x)$ is a Broyden function in [18]. That is,

$$\begin{aligned} y_1 &= (3 - 2x_1)x_1 - 2x_2 + 1, \\ y_i &= (3 - 2x_i)x_i - x_{i-1} - 2x_{i+1} + 1, \quad i = 2, 3, \ldots, n-1 \\ y_n &= (3 - 2x_n)x_n - x_{n-1} + 1, \end{aligned}$$

and *inner-p* can be specified by users to control the workload for each basic function $g(x)$. In our examples, we take *inner-p* = 50. We define such a basic function $g(x)$ to make the workload for $F_i(x)$ heavy enough to justify the use of an intermediate variable y_i. The Brown function in [78] is adopted for function $\bar{F}(x, y_1, \ldots, y_p)$. That is,

$$\begin{aligned} y_i &= (x_i^2)^{x_{i+1}^2+1}, \qquad i = 1, \ldots, n-1; \\ y_i &= y_i + (x_{i+1}^2)^{x_i^2+1}, \qquad i = 1, \ldots, n-1; \\ z &= \textstyle\sum_{i=1}^{n}(y_i). \end{aligned}$$

Using the basic function $g(x)$ and $\bar{F}(x, y_1, \ldots, y_p)$, we can construct the dynamic system and generalize partial separability examples.

Example 4.3. (Jacobian computation for the three-step function) Consider the composite function, $F(x) = \bar{F}(A^{-1}\tilde{F}(x))$, where $\bar{F}$ and $\tilde{F}$ are Broyden functions [17] (their Jacobian matrices are tridiagonal) and the structure of A is based on 5-point Laplacian defined on a square $(\sqrt{n}+2)$-by-$(\sqrt{n}+2)$ grid. For each nonzero element of A, A_{ij} is defined as a function of x, specifically, $A_{ij}=x_j$. Thus, the nonzero elements of A depend on x; the structure of $A_x \cdot v$, for any v, is equal to the structure of A.

The evaluation of $z = F(x)$ is a structured computation, $F^E(x,y_1,y_2)$, defined by the following three steps:

$$\left.\begin{array}{lll} (1) \text{ Solve for } y_1 & : & y_1-\tilde{F}(x)=0 \\ (2) \text{ Solve for } y_2 & : & Ay_2-y_1=0 \\ (3) \text{ Solve for } z & : & z-\bar{F}(y_2)=0 \end{array}\right\}. \tag{4.24}$$

Differentiating F^E defined by (4.24) with respect to x, y_1, y_2 yields

$$J^E = \begin{bmatrix} -\tilde{J} & I & 0 \\ A_x y_2 & -I & A \\ 0 & 0 & \bar{J} \end{bmatrix}. \tag{4.25}$$

■

The Jacobian matrix J can be computed via (4.2). The main MATLAB code using ADMAT for this example can be listed as follows:

```
%
% get all yi for the structure Jacobian construction
%
   y = zeros(3*n,1);
   y(1:n,1) = x;
   y(n+1:2*n,1) = broyden(y(1:n,1));
   y(2*n+1:3*n,1) = F2(y(1:2*n,1), Extra);
%
% Structured Jacobian construction
%
% Compute Jbar via the reverse mode.
%
   myfun1 = ADfun('barF', n);
   options = setopt('revprod',eye(n));
   [z, Jbar] = feval(myfun1, y, Extra, options);
%
% Solve the expanded block linear system by the reverse mode
%
   myfun2 = ADfun('F2', n);
   options = setopt('revprod', Jbar(2*n+1:3*n,:));
   [t, tmp] = feval(myfun2,y(1:2*n,1), Extra, options);
   Jbar(1:2*n,:) = Jbar(1:2*n,:) + tmp;
   myfun2 = ADfun('broyden', n);
   options = setopt('revprod', Jbar(n+1:2*n,:));
   [t1, tmp] = feval(myfun2,y(1:n,1), Extra, options);
   Jbar(1:n,:) = Jbar(1:n,:) + tmp;
% get the Jacobian matrix of the original function
   J = Jbar(1:n,:);
```

Table 4.1. *Memory usage and running times of the gradient computation based on the structure and plain reverse-mode AD for the generalized partial separability problem.*

	Reverse mode		Structured gradient		Memory	Time
p	Memory (MB)	Time (s)	Memory (MB)	Time (s)	Ratio	Speedup
10	48.55	4.86	13.55	3.08	3.58	1.58
30	214.73	14.66	13.58	9.96	15.81	1.47
50	473.07	24.54	13.60	14.95	34.78	1.64
100	1232.21	52.18	13.68	29.95	90.00	1.74
150	2547.30	837.77	13.72	57.56	185.66	16.06

Example 4.4. (Gradient of a generalized partial separable function) Suppose the function can be defined as

$$\begin{array}{ll} \text{for} & i=1,\ldots,p \\ & \text{Solve for } y_i : g(x)-y_i=0. \\ \text{and} & \text{then} \\ & \text{Solve for } z : \bar{F}(x,y_1,\ldots,y_p)-z=0, \end{array}$$

where $g(x)$ is the function defined as the Broyden function and $\bar{F}(x,y_1,\ldots,y_p)$ is the Brown function. In this case the auxiliary Jacobian matrix structure (4.6) reduces to

$$J^E = \left(\begin{array}{c|cccc} -J_1 & I & & & \\ -J_2 & & I & & \\ \vdots & & & \ddots & \\ -J_p & & & & I \\ \hline 0 & \bar{J}_1 & \bar{J}_2 & \cdots & \bar{J}_p \end{array}\right), \tag{4.26}$$

where $\bar{J}$ is divided into different components $\bar{J}_i$ corresponding to variable vector y_i and J_i is the Jacobian of g with respect to x at each iteration of the for loop, $i=1,\ldots,p$. ■

The main MATLAB code using ADMAT to compute the gradient for this example can be programmed as follows. Table 4.1 lists all computational times and memory required by the structured ideas and the reverse mode. It shows that the structured idea can reduce the computational times and memory requirement significantly.

```
% get all y_i for the structure Jacobian construction
%
   y = zeros((p+1)*m1,1);
   y(1:m1,:)  = x; % where m1 is the length of y_i
   for i = 1 :  p
   y(i*m1+1:   (i+1)*m1) = feval(TildeFun, x, Extra);
   end
%
% Structured Jacobian construction
%
% Compute Jbar, via the reverse mode AD
%
   myfun1 = ADfun(Fbar, (p+1)*m1);
```

```
    options = setopt('revprod', eye(m2));
    [ z, Jbar ] = feval(myfun1, y, Extra, options);
    Jbar = Jbar(:);
%
% solve the expanded block linear system
%
    myfun2 = ADfun(TildeFun, m1);
    for i = p:-1:1
    options.W = Jbar(m1*i+1:m1*(i+1),:);
    [t, tmp] = feval(myfun2, x, Extra, options);
    Jbar(1:m1,:)  = Jbar(1:m1,:)+tmp(:);
    end
% Got the Jacobian matrix of the original function
    J = Jbar(1:m1,:);
```

The dynamic system, a special case of a composite function, represents another extreme case for the structured gradient computation. In the GPS structure, the intermediate variable y_i depends only on the independent variable x, not other intermediate variables, whereas in the dynamic system, the intermediate variable y_i depends only on the previous intermediate variable y_{i-1}, not others.

Example 4.5. (Gradient of a composite function) Suppose the function has the form

$$\begin{array}{ll} \text{for} & i=1,2,\ldots,p \\ & \text{Solve for } y_i : g(y_{i-1})-y_i=0, \\ \text{and} & \text{then} \\ & \text{Solve for } z : \bar{F}(x,y_1,\ldots,y_p)-z=0, \end{array} \tag{4.27}$$

where $y_0=x$, $g(x)$ is the function defined as the Broyden function and $\bar{F}(x,y_1,\ldots,y_p)$ is the Brown function. The corresponding extended Jacobian matrix is

$$J^E = \begin{pmatrix} J_x^1 & -I & & & \\ 0 & J_{y_1}^2 & -I & & \\ \vdots & \vdots & \ddots & \ddots & \\ 0 & 0 & \vdots & J_{y_{p-1}}^p & -I \\ \bar{J}_x & \bar{J}_{y-1} & \cdots & \cdots & \bar{J}_{y_p} \end{pmatrix}.$$

■

The gradient of this composition function can be programmed via MATLAB using the following ADMAT toolbox. Results in Table 4.2 show that the structure idea reduces the memory usage significantly, especially when the number of recursions, p, is large; i.e., the original function evaluation is expensive. As p increases, the memory usage by the plain reverse-mode AD increases dynamically, while the structure gradient computation is mild in the memory usage. Our structure method has the same order of computational cost as direct reverse mode. However, due to the saving in memory, our method reduces the memory accessing time; thus, it leads to a smaller computational time than the plain reverse-mode AD.

Table 4.2. *Memory usage and running times of the gradient computation based on the structure and plain reverse-mode AD for the dynamic system.*

	Reverse mode		Structured gradient		Memory	Time
p	Memory (MB)	Time (s)	Memory (MB)	Time (s)	Ratio	Speedup
10	29.27	3.75	3.56	3.15	8.22	1.19
30	87.71	10.27	3.62	8.96	24.23	1.15
50	146.17	16.98	3.68	15.05	39.72	1.13
100	292.30	33.76	3.85	29.73	75.92	1.14
150	438.43	51.00	3.99	45.25	109.88	1.13

```
% get all yi for the structure Jacobian construction
   y = zeros((p+1)*m1,1);
   y(1:m1,:)  = x;
   y(m1+1:2*m1,:)  = feval(fun, x, Extra);
   for i = 2 :  p
       Extra.start = (i-1)*m1+1;
       Extra.end = i*m1;
       y(i*m1+1:  (i+1)*m1) = feval(fun, y, Extra);
   end
%
% Structured Jacobian construction
%
% Compute Jbar, the right-hand side of the linear system.
   Extra.start = p*m1+1;
   Extra.end = (p+1)*m1;
   myfun1 = ADfun(Fbar, (p+1)*m1);
   options = setopt('revprod', eye(m2));
   [z, Jbar] = feval(myfun1, y,Extra, options);
   Jbar = (Jbar(:));
%
% solve the block linear system
   myfun2 = ADfun(fun, m1);
   for i = p:-1:1
       Extra.start = (i-1)*m1+1;
       Extra.end = i*m1;
       options = setopt('revprod', Jbar(m1*i+1:m1*(i+1),:));
       [t, tmp] = feval(myfun2, y, Extra, options);
       tmp = (tmp(:));
       Jbar(1:i*m1,:)  = Jbar(1:i*m1,:)+tmp(1:i*m1,:);
   end
   J = Jbar(1:m1,:);
```

4.5 ▪ Structured Hessian Computation

We now consider a structured scalar-valued function $f(x) : \mathcal{R}^n \rightarrow \mathcal{R}$ and the determination of the Hessian matrix based on the method [32]. First recall from Chapter 2 that AD can be applied to function f, assuming no sparsity or structure, to obtain the Hessian matrix $H(x) = \nabla^2 f(x)$, in time proportional to $n \cdot \omega(f)$ and space $\omega(f)$. If the Hessian matrix H is sparse, then the running time is proportional to $\chi(H)\cdot\omega(f)$, where $\chi(H)$ is a measure of the sparsity in H (corresponds to a cyclic chromatic number of the adjacency graph of H; see [31]) and the space required is proportional to $\omega(f)$. However, in reality, many expensive functions f do not exhibit sparse Hessian matrices, but they are structured.

The general structure we consider, similar to that given in the previous section, is the following. To evaluate $z=f(x):\mathcal{R}^n\to\mathcal{R}$,

$$\left.\begin{array}{l}
\text{Solve for } y_1: \ F_1(x,y_1)\equiv \tilde{F}_1(x)-M_1\cdot y_1=0\\
\text{Solve for } y_2: \ F_2(x,y_1,y_2)\equiv \tilde{F}_2(x,y_1)-M_2\cdot y_2=0\\
\vdots\\
\text{Solve for } y_p: \ F_p(x,y_1,y_2,\dots,y_p)\equiv \tilde{F}_p(x,y_1,y_2,\dots,y_{p-1})-M_p\cdot y_p=0\\
\text{Solve for } z: \ F_{p+1}(x,y_1,y_2,\dots,y_{p+1})\equiv \bar{f}(x,y_1,y_2,\dots,y_p)-z=0
\end{array}\right\}, \tag{4.28}$$

i.e., solve $F^E(x,y)=0$, where the $\tilde{F}_i$'s $(i=1,\dots,p)$ and $\bar{f}$ are intermediate functions and each matrix M_i is square and nonsingular. The intermediate vectors y_i's are of varying dimension. Denote the functions in (4.28) as

$$F^E(x,y)=\begin{bmatrix}\tilde{F}_1(x)-M_1y_1\\ \tilde{F}_2(x,y_1)-M_2y_2\\ \vdots\\ \tilde{F}_p(x,y_1,\dots,y_{p-1})-M_py_p\\ \bar{f}(x,y_1,\dots,y_p)\end{bmatrix}. \tag{4.29}$$

The corresponding *extended* Jacobian matrix of F^E comes from differentiating (4.29) with respect to (x,y) and can be written as

$$J^E=\begin{pmatrix} J_x^1 & -M_1 & & & \\ J_x^2 & J_{y_1}^2 & -M_2 & & \\ \vdots & \vdots & \vdots & \ddots & \\ J_x^p & J_{y_1}^p & \vdots & J_{y_{p-1}}^p & -M_p\\ \nabla\bar{f}_x^T & \nabla\bar{f}_{y_1}^T & \cdots & \cdots & \nabla\bar{f}_{y_p}^T\end{pmatrix}.$$

If we partition J^E as

$$J^E=\left(\begin{array}{c|cccc} J_x^1 & -M_1 & & & \\ J_x^2 & J_{y_1}^2 & -M_2 & & \\ \vdots & \vdots & \vdots & \ddots & \\ J_x^p & J_{y_1}^p & \vdots & J_{y_{p-1}}^p & -M_p\\ \hline \nabla\bar{f}_x^T & \nabla\bar{f}_{y_1}^T & \cdots & \cdots & \nabla\bar{f}_{y_p}^T\end{array}\right)=\left(\begin{array}{c|c}\hat{J}_x^E & \hat{J}_y^E\\ \hline \nabla\bar{f}_x^T & \nabla\bar{f}_y^T\end{array}\right), \tag{4.30}$$

then using a Schur-complement computation, the gradient of f satisfies

$$\nabla f^T=\nabla\bar{f}_x^T-\nabla\bar{f}_y^T(\hat{J}_y^E)^{-1}\hat{J}_x^E. \tag{4.31}$$

Recall that Algorithm 4.1 is an efficient method to compute (4.30) and does not require the explicit computation of submatrices $J_{y_i}^i$, $i > 2$, $j = 1,\ldots,p-1$.

The Hessian can be calculated with the following structured approach. Define a function,

$$g(x,y,w) = \bar{f}(x,y) + \sum_{i=1}^{p} w_i^T F_i(x,y), \tag{4.32}$$

where w_i is solution of $(\hat{J}_{y_i}^E)^T w = -\nabla \bar{f}_{y_i}$. Now define the *auxiliary* Hessian matrix, which is obtained by using $(\hat{J}_y^E)$ in (4.30) and differentiating on $g(x,y,w)$ in (4.32) with respect to (x,y,w):

$$H^a = \begin{bmatrix} \hat{J}_x^E & \hat{J}_y^E & 0 \\ \nabla_{yx}^2 g & \nabla_{yy}^2 g & (\hat{J}_y^E)^T \\ \nabla_{xx}^2 g & \nabla_{xy}^2 g & (\hat{J}_x^E)^T \end{bmatrix}.$$

Partition H^a as

$$H^a = \left[\begin{array}{c|cc} \hat{J}_x^E & \hat{J}_y^E & 0 \\ \nabla_{yx}^2 g & \nabla_{yy}^2 g & (\hat{J}_y^E)^T \\ \hline \nabla_{xx}^2 g & \nabla_{xy}^2 g & (\hat{J}_x^E)^T \end{array}\right] = \left[\begin{array}{c|c} A & L \\ \hline B & M \end{array}\right].$$

The Hessian matrix of f is related to H^a by a Schur-complement computation:

$$H = B - ML^{-1}A. \tag{4.33}$$

Expression (4.33) indicates how to compute the Hessian matrix given the component pieces in (4.32). However, there are different ways to implement (4.33), with different costs, and we detail a particularly efficient approach below.

Note that we can write

$$L = \begin{bmatrix} \hat{J}_y^E & 0 \\ \nabla_{yy}^2 g & (\hat{J}_y^E)^T \end{bmatrix} = \begin{bmatrix} I & 0 \\ \nabla_{yy}^2 g(\hat{J}_y^E)^{-1} & (\hat{J}_y^E)^T \end{bmatrix} \cdot \begin{bmatrix} \hat{J}_y^E & 0 \\ 0 & I \end{bmatrix},$$

and both matrices in the product on the right-hand side are nonsingular. Therefore,

$$L^{-1} = \begin{bmatrix} \hat{J}_y^E & 0 \\ 0 & I \end{bmatrix}^{-1} \begin{bmatrix} I & 0 \\ \nabla_{yy}^2 g(\hat{J}_y^E)^{-1} & (\hat{J}_y^E)^T \end{bmatrix}^{-1} \tag{4.34}$$

and

$$ML^{-1}A = \left(M \begin{bmatrix} \hat{J}_y^E & 0 \\ 0 & I \end{bmatrix}^{-1} \right) \cdot \left(\begin{bmatrix} I & 0 \\ \nabla_{yy}^2 g(\hat{J}_y^E)^{-1} & (\hat{J}_y^E)^T \end{bmatrix}^{-1} A \right).$$

Then using (4.34),

$$\begin{aligned} H &= B - ML^{-1}A \\ &= \nabla_{xx}^2 g - (\nabla_{xy}^2 g \cdot (\hat{J}_y^E)^{-1}) \cdot \hat{J}_x^E - [(\hat{J}_x^E)^T (\hat{J}_y^E)^{-T} \cdot [\nabla_{yx}^2 g - (\nabla_{yy}^2 g)(\hat{J}_y^E)^{-1} \hat{J}_x^E]]. \end{aligned} \tag{4.35}$$

We now describe how to compute (4.33) in more detail.

4.5.1 ▪ An Explicit Method

First we note that for any matrix

$$V = \begin{bmatrix} V_1 \\ \cdots \\ V_p \end{bmatrix}$$

with the number rows of V_i equal to the order of matrix M_i, $i = 1, \ldots, p$, the product $W = (\hat{J}_y^E)^{-T} V$ can be computed by Algorithm 4.2.

Algorithm 4.2.
Explicit-Inverse-Product (EIP) to compute $W = (\hat{J}_y^E)^{-T} V$

1. Following Algorithm 4.1, evaluate $y_1, y_2, \ldots, y_p$.
2. (a) Initialize: $T_i = 0$, $i = 1, \ldots, p$.
   (b) For $j = p, p-1, \ldots, 1$
   Solve $M_j W_j^T = T_j - V_j^T$,
   Set $T_i = T_i + W_j^T J_{y_i}^j$ $(i = 1, \ldots, j-1)$.

This leads to our explicit Hessian computation algorithm as follows.

Algorithm 4.3.
Explicit-Compute-Hessian (ECH)

1. Evaluate $y_1, y_2, \ldots, y_p$;
2. For $i = 1, 2, \ldots, p$
   Evaluate $F_i(x)$, $F_i(y_j)$ $j = 1, \ldots, i-1$.
   Compute M_i, $J_{y_j}^i$, J_x^i $(j = 1, \ldots, i-1)$ using sparse AD.
3. Solve $(\hat{J}_y^E)^T w = -\nabla_y \bar{f}$ to obtain w, using Algorithm 4.1.
4. Let $g(x,y) = \bar{f}(x,y) + \sum_{i=1}^p w_i^T F_i(x,y)$ and compute its Hessian,
   $$\nabla^2 g = \begin{bmatrix} \nabla_{xx}^2 g & \nabla_{yx}^2 g \\ \nabla_{xy}^2 g & \nabla_{yy}^2 g \end{bmatrix} = \nabla^2 \bar{f} + \sum_{i=1}^p \nabla^2 (w_i^T F_i),$$
   using reverse-mode sparse AD.
5. Using Algorithm 4.2, compute $R^T = (\bar{J}_y^E)^{-T} (\nabla_{xy}^2 g)^T$ and $C^T = (\hat{J}_y^E)^{-T} (\nabla_{yy}^2 g)^T$.
6. Compute $T = C \cdot \hat{J}_x^E - C_1 J_x^1 + \cdots + C_p J_x^p$.
7. Using Algorithm 4.2, compute $S = (\hat{J}_x^E)^T (\hat{J}_y^E)^{-T} [\nabla_{yx}^2 g - T]$.
8. Set $\nabla^2 f = \nabla_{xx}^2 g - \sum_{i=1}^p R_i J_x^i - \sum_{i=1}^p S_i^T J_x^i$.

The total time and space requirements for ECH are

$$\begin{aligned}\omega &\sim \textstyle\sum_{i=1}^{p}\left[\chi(J_x^i)+\sum_{j=1}^{i}\chi(J_{y_j}^i)+\chi(\nabla^2(w_i^T F_i))\right]\cdot\omega(F),\\ \sigma &\sim \sigma(J_{y_j}^i)+\max_i\{\omega(F_i),\ \omega(\bar{f})\}.\end{aligned}$$

Comparing the costs for ECH to that of unstructured sparse AD in [104], it is unclear whether ECH has lower time complexity: execution time depends on the sparsity of all the functions involved. However, as we illustrate below, the computing time can be considerably less than the unstructured approach. Moreover, ECH is much more efficient memory-wise.

4.5.2 ▪ Examples

We consider three test example classes for the structured Hessian computation. There are two extreme cases: the dynamic system and the generalized partially separable problem. In the dynamic system, each intermediate variable depends only on the intermediate variable preceding it, while in the general partially separable problem, each intermediate variable depends only on the input variable x. The third problem represents a mix of these two extreme cases.

I. Dynamic System

The dynamic system $z=f(x)$ is defined as follows:

$$\begin{array}{ll} y_0 = x & \\ \text{for} & i=1,2,\ldots,p \\ & \text{Solve for } y_i : F_i(y_{i-1})-y_i=0, \\ \text{and} & \text{then} \\ & \text{Solve for } z : \bar{f}(x,y_1,\ldots,y_p)-z=0. \end{array} \tag{4.36}$$

For our experiments, the intermediate function S is defined as the Broyden function [18]:

$$\begin{array}{ll} \text{for} & j=1:inner-p \\ & y_1=(3-2v_1)v_1-2v_2+1, \\ & y_i=(3-2v_i)v_i-v_{i-1}-2v_{i+1}+1, \quad i=2,3,\ldots,n-1 \\ & y_n=(3-2v_n)v_n-v_{n-1}+1, \\ \text{end,} & \end{array} \tag{4.37}$$

where $inner-p$ is a variable allowing us to control the workload of the intermediate functions. The last intermediate function $\bar{f}$ is defined as the Brown function,

$$\begin{array}{ll} y_i=(v_i^2)^{v_{i+1}^2+1}, & i=1,\ldots,n-1; \\ y_i=y_i+(v_{i+1}^2)^{v_i^2+1}, & i=1,\ldots,n-1; \\ z=\sum_{i=1}^{n}(y_i). & \end{array} \tag{4.38}$$

z is dependent only on the last intermediate variable, y_p.

II. Generalized Partial Separability

The generalized partially separable function $z = f(x)$ is defined as follows:

$$\begin{array}{ll} \text{for} & i = 1,\ldots,p \\ & \text{Solve for } y_i : F_i(x) - y_i = 0. \\ \text{and} & \text{then} \\ & \text{Solve for } z : \bar{f}(x, y_1, \ldots, y_p) - z = 0. \end{array} \tag{4.39}$$

For our experiments, each intermediate function F_i is defined as the Broyden function in (4.37), and $\bar{f}$ is defined as the Brown function in (4.38).

III. A General Case

We also look at a more general case that lies between the dynamic system and generalized partial separability cases, where $z = f(x)$ is defined as

$$\left.\begin{array}{l} \text{Solve for } y_i : \; y_i - T_i(x) = 0 \quad i = 1,\ldots,6 \\ \text{Solve for } y_7 : \; y_7 - T_7((y_1 + y_2)/2) = 0 \\ \text{Solve for } y_8 : \; y_8 - T_8((y_2 + y_3 + y_4)/3) = 0 \\ \text{Solve for } y_9 : \; y_9 - T_9((y_5 + y_6)/2) = 0 \\ \text{Solve for } y_{10} : \; y_{10} - T_{10}((y_7 + y_8)/2) = 0 \\ \text{Solve for } y_{11} : \; y_{11} - T_{11}((y_8 + y_9)/2) = 0 \\ \text{Solve for } z : \; z - 0.4(y_{10} + y_{11}) - 0.2 y_5 = 0 \end{array}\right\}, \tag{4.40}$$

where $T_i(x)$ $(i = 1, 2, \ldots, 6)$ are defined as $g(x)$, and $T_i(x)$ $(i = 7, \ldots, 11)$ are defined as

$$\begin{array}{l} y(1) = (3 - 2 * x(1)).* x(1) - 2 * x(2) + 1; \\ y(i) = (3 - 2 * x(i)).* x(i) - x(i-1) - 2 * x(i+1) + ones(n-2, 1); \\ y(n) = (3 - 2 * x(n)).* x(n) - x(n-1) + 1 \end{array}$$

for $i = 2, \ldots, n-1$. It is clear that the computation of $T_i(x)$ $(i = 1, \ldots, 6)$ dominates the running time of evaluating $z = F(x)$. The structure of the corresponding expanded Jacobian matrix is illustrated as

$$J^E = \left[\begin{array}{cccccccccccc}
X & X & & & & & & & & & & \\
X & & X & & & & & & & & & \\
X & & & X & & & & & & & & \\
X & & & & X & & & & & & & \\
X & & & & & X & & & & & & \\
X & & & & & & X & & & & & \\
 & X & X & & & & & X & & & & \\
 & & X & X & X & & & & X & & & \\
 & & & & & X & X & & & X & & \\
 & & & & & & & X & X & & X & \\
 & & & & & & & & X & X & & X \\
 & & & & & X & & & & & X & X
\end{array}\right],$$

where X represents the nonzero block.

Chapter 5

Newton's Method and Optimization

Nonlinear minimization and solving nonlinear systems of equations often involve the computation of the Newton step. For a nonlinear system, i.e., find $x \in \mathcal{R}^n$ such that $F(x) = 0$, where $F : \mathcal{R}^n \to \mathcal{R}^n$ and continuously differentiable, the determination of the Newton step typically involves two major steps:

$$\left.\begin{array}{lll} 1. & \text{Evaluate } F(x), & J(x) \in \mathcal{R}^{n\times n} \\ 2. & \text{Solve} & J \cdot s_N = -F \end{array}\right\}, \tag{5.1}$$

where s_N is the Newton step. The Newton step for nonlinear minimization,

$$\min_x f(x), \qquad \text{where } f : \mathcal{R}^n \to \mathcal{R}^1$$

and f is twice continuously differentiable, is similar to (5.1) with $J(x)$ replaced by the Hessian matrix of $f(x)$ and with $F(x)$ replaced by the gradient $\nabla f(x)$. Below we focus on the nonlinear system case, $F(x) = 0$, and return to nonlinear minimization in Section 5.2.

5.1 ▪ Newton Step for Nonlinear Systems

The two steps in (5.1) are usually the most expensive computations in the user code, with step 1 often dominant. However, if function F is structured as defined in (4.5), then it may be possible to compute the Newton step s_N without fully forming the Jacobian matrix $J(x)$, using the AD technology discussed in Chapter 4. In many cases this allows for the use of sparse AD methods (even when $J(x)$ is dense) and can result in significant efficiency gains. To see this, consider the form in (4.5) and the corresponding extended Jacobian matrix J^E in (4.6). Recall that each component function in (4.5), i.e., F_i^E, can be evaluated and differentiated separately, in a top-to-bottom order. Moreover, if the Jacobian of F_i^E is sparse, i.e., J_i^E is sparse, then sparse AD techniques can be used, strip by strip, to get the entire extended Jacobian matrix J^E. Consequently, the work required to obtain J^E satisfies $\omega(J^E) \leq \chi(J^E) \cdot \omega(F)$ (see (4.7)), and the space requirements are $\sigma(J^E) \sim \max\{\omega(F_i^E),\ i = 1, \ldots, p+1\}$.

Next, to compute the Newton step, we must solve the large sparse block system

$$J^E \begin{bmatrix} s_N \\ d \end{bmatrix} = \begin{bmatrix} 0 \\ -F(x) \end{bmatrix}, \tag{5.2}$$

where J^E is given in (4.6). One solution, as indicated by (4.8), is the Schur complement computation. However, this is just one way to solve (5.2); there are other ways, including a general sparse solver applied to (5.2). The most effective method will depend on the specific block/sparsity structure of the extended Jacobian J^E. The results given in [44] indicate that if F is structured and J is dense, then it is *significantly* more efficient to get the Newton step by obtaining J^E using the sparse AD techniques and then solve (5.2) by a block/sparse solver of some sort as opposed to obtaining J directly by AD (or finite differences).

For example, consider the autonomous ODE,

$$y' = f(y).$$

If $y(0) = f(x_0)$, then we use an explicit one-step Euler method to compute an approximation y_k to a desired final state $y(T) = \phi(u_0)$. Thus, the one-step Euler method leads to a recursive function,

$$\begin{array}{l} y_0 = x \\ \text{for } i = 1, \ldots, p \\ \qquad \text{Solve for } y_i : \; y_i - F(y_{i-1}) = 0 \\ \text{Solve for } z : \; z - y_p = 0, \end{array} \tag{5.3}$$

where $F(y_i) = y_i + h \cdot f(y_i)$ and h is the step size of the Euler method. The corresponding expanded function is

$$F^E(x, y_1, \ldots, y_p) = \begin{pmatrix} y_1 - F(y_0) \\ y_2 - F(y_1) \\ \vdots \\ y_p - F(y_{p-1}) \\ y_p \end{pmatrix}.$$

The subsequent Newton process can be written as

$$J^E \begin{pmatrix} \delta x \\ \delta y_1 \\ \delta y_2 \\ \vdots \\ \delta y_p \end{pmatrix} = -F^E = \begin{pmatrix} 0 \\ 0 \\ 0 \\ \vdots \\ -G \end{pmatrix}, \tag{5.4}$$

where J^E is the Jacobian matrix of $F^E(x, y_1, \ldots, y_p)$ with respect to $(x, y_1, \ldots, y_p)$ and G is the function in the nonlinear equation $G(x) = 0$. In particular, $G(x) = y - \phi(u_0)$ in this example. Following is the main MATLAB code for implementing the expanded Newton step (5.4). Note that in our implementation, the sparse Jacobian computation, `evalj`, is employed, instead of the plain forward and reverse mode of AD, due to the sparsity of each block in J^E.

```
[x,fval, it] = newton_expanded(func, func_y, funG, x0, tol, itNum, Extra, p)
        ⋮
        ⋮
x = x0;
m = length(x0); % size of x
% compute [x; y1; ...; yp]
y = feval(func_y, x, Extra);
n = length(y); % size of [x,y1,y2,...,yp]

% initialize the number of iteration steps
it = 0;
num_int = 20;
if isempty(Extra)
    % get the sparse structure of Jacobian matrix of func
    JPI = getjpi(func,n);
else    % Extra is not empty
    % get the sparse structure of Jacobian matrix of func
    JPI = getjpi(func, n,[], Extra);
end

fval= 1;
% Newton steps
while (abs(norm(fval)) > tol) && (it < itNum)
    % calculate the function value of G(x).
    fval = feval(funG, y, Extra);
    k = length(fval);
    % evaluate the function value and the Jacobian matrix at [x,y1,...,yp]
    [z, J] = evalj(func, y, Extra, [],JPI);

    % if number of intermediate variables is larger than num_int,
    % the speed of the structured Newton method is low since the
    % overload on solving a linear Newton system.  Thus, we absorb
    % some intermediate variables in order to decrease the size
    % of the Newton system.
    if p > num_int
        N = n/(p+1); % size of original problem
        for i = p:-1:  num_int
            D = J(i*N+1:(i+1)*N, i*N+1:(i+1)*N);
            B = J((i-1)*N+1:i*N, i*N+1:(i+1)*N);
            A = J((i-1)*N+1:i*N, 1:i*N);
            C = J(i*N+1:(i+1)*N, 1 :i*N);
            C = C-D*(B\ A);
            J = sparse([J(1:(i-1)*N, 1:i*N); C]);
        end
        % construct right hand side of Newton system
        F = zeros(num_int*N,1);
        F(N*(num_int-1)+1 :  num_int*N) = fval;
    else
    % construct right hand side of Newton system
        F = zeros(n,1);
        F(n-k+1 :  n) = fval;
    end
    delta = -J\ F;
    x = x + delta(1:m);
    % compute [x; y1; ...; yp]
    y = feval(func_y, x, Extra);
    it = it + 1;
end
```

The following example shows how to solve a specific ODE via the expanded Newton method.

Example 5.1. Solve the ODE via the expanded Newton method.

See DemoNewton_Exp.m

1. Set some initial values.

```
>> p = 5;   % number of intermediate variables
>> N = 8;   % size of original problem
>> tol = 1e-13;   % convergence tolerance
>> itNum = 40;   % maximum number of iterations
>> h = 1e-8;   % step size
```

2. Initialize the random seed.

```
>> rand('seed', 0);
```

3. Set the target function's input point xT.

```
>> xT = rand(N,1)
xT =
    0.2190
    0.0470
    0.6789
    0.6793
    0.9347
    0.3835
    0.5194
    0.8310
```

4. Set the expanded function F^E, F, and G.

```
>> func = 'Exp_DS';   % Expanded function with independent
>>                     % variables, x, y1, y2,..., yM
>> func_y = 'func_DS';   % function revealing relation between
>>                        % x, y1,y2,...,yM
>> funG = 'Gx';   % right hand side of Newton process
```

5. Set parameters required in func = 'Exp_DS' and func_y.

```
>> Extra.N = N;   % size of original problem
>> Extra.M = p;   % number of intermediate variables
>> Extra.u0 = xT;   % input variable for the target function φ(xT)
>> Extra.fkt = @fx;   % function f on right hand side of ODE
>> Extra.phi = @exp;   % target function φ
>> Extra.h = h;   % step size of one step Euler method.
```

Note that we set the right-hand-side function of the ODE to be $f(x) = x$ so that the actual solution of the above ODE is $y = e^x$.

6. Set the starting point x.

```
>> x = ones(n,1);
```

7. Solve the ODE by the expanded Newton computation.

```
>> [x, fval, it] = newton_expand(func, func_y, x, tol, itNum, Extra, p)
x =

    1.2448
    1.0482
    1.9716
    1.9725
    2.5464
    1.4674
    1.6810
    2.2955

fval =

    1.0415e-015*

    0
    0
    0.2220
    0
    0.4441
    0.2220
    0
    0.8882

it =
    2
```

8. Compare the computed solution with the target function value.

```
>> Difference between computed solution with target function value
norm(x-exp(xT)) = 2.605725e-007
```

■

5.2 ▪ Newton Step for Nonlinear Minimization

The Newton step for nonlinear minimization typically involves two major steps:

$$\left.\begin{array}{lll} 1. & \text{Evaluate } f(x),\ \nabla f(x)\in\mathscr{R}^n, & H(x)\equiv\nabla^2 f(x)\in\mathscr{R}^{n\times n} \\ 2. & \text{Solve} & H\cdot s_N=-\nabla f(x) \end{array}\right\}. \tag{5.5}$$

Step 1 in (5.5) is often the most expensive part of the algorithm, as it involves the evaluation of $H(x)$. When $H(x)$ is dense, the computing cost to execute this step, using AD, is proportional to $n\cdot\omega(f)$ in space $\sigma\sim\omega(f)$. Alternatively, if only forward mode is used, the computational time is proportional to $n^2 \cdot \omega(f)$ in space $\sim \sigma(f)$.[5] However, if function f is structured as given in (4.21), then it is possible to do much better.

Assume the structure given in (4.21); then the main steps to evaluate $z = f(x) : \mathcal{R}^n \to \mathcal{R}$ are

$$\left.\begin{array}{l} \text{Solve for } y_1 : \; F_1(x, y_1) \equiv \tilde{F}_1(x) - M_1 \cdot y_1 = 0 \\ \text{Solve for } y_2 : \; F_2(x, y_1, y_2) \equiv \tilde{F}_2(x, y_1) - M_2 \cdot y_2 = 0 \\ \vdots \\ \text{Solve for } y_p : \; F_p(x, y_1, y_2, \ldots, y_p) \equiv \tilde{F}_p(x, y_1, y_2, \ldots, y_{p-1}) - M_p \cdot y_p = 0 \\ \text{Solve for } z : \; F_{p+1}(x, y_1, y_2, \ldots, y_{p+1}) \equiv \bar{f}(x, y_1, y_2, \ldots, y_p) - z = 0. \end{array}\right\}, \qquad (5.6)$$

i.e., solve $F^E(x, y) = 0$, where the $\tilde{F}_i$'s $(i = 1, \ldots, p)$ and $\bar{f}$ are intermediate functions and each matrix M_i is square and nonsingular. The Newton step s_N given in (5.5) is also the solution to the augmented system,

$$\begin{bmatrix} \hat{J}_x^E & \hat{J}_y^E & 0 \\ \nabla_{xy}^2 g & \nabla_{yy}^2 g & (\hat{J}_y^E)^T \\ \nabla_{xx}^2 g & \nabla_{xy}^2 g & (\hat{J}_x^E)^T \end{bmatrix} \begin{bmatrix} s_N \\ d_1 \\ d_2 \end{bmatrix} = - \begin{bmatrix} 0 \\ 0 \\ \nabla f \end{bmatrix}, \qquad (5.7)$$

where $g(x, y) = \bar{f}(x, y) + \sum_{i=1}^p w_i^T F_i(x, y)$ and w_i solves $(\hat{J}_{y_i}^E)^T w = -\nabla \bar{f}_{y_i}$ as defined in (4.30). There are a number of ways to solve (5.7), including forming H by the procedure in Section 4.5 and then solving $H s_N = -\nabla f(x)$. However, a sparse-block technique applied directly to (5.7) can sometimes be more efficient.

Note that the work required to obtain

$$H^a = \begin{bmatrix} \hat{J}_x^E & \hat{J}_y^E & 0 \\ \nabla_{xy}^2 g & \nabla_{yy}^2 g & (\hat{J}_y^E)^T \\ \nabla_{xx}^2 g & \nabla_{xy}^2 g & (\hat{J}_x^E)^T \end{bmatrix}$$

is proportional to $\chi(H^a) \cdot \omega(f)$, where $\chi(H^a)$ is a measure of the sparsity in H^a (see Chapter 3). Therefore, the total time to compute the Newton step s_N by this procedure is $\chi(H^a) \cdot \omega(f)$ plus the time to do the sparse linear solve (5.7). Next, we give an example to show how to use the expanded Hessian matrix to implement one Newton step for a minimization problem.

Example 5.2. Structured Newton computation for solving a minimization problem.

See DemoExpHess.m ∎

We minimize $z = \bar{F}(\tilde{F}(x)) + x^T x$, where $\tilde{F}(x)$ is the Broyden function and $\bar{F}(y)$ is a scalar-valued function, $\bar{F}(y) = \sum_{i=1}^n y_i^2 + \sum_{i=1}^{n-1} 5 y_i y_{i+1}$. The triangular computation system can be constructed as follows:

$$\begin{array}{lcl} \text{Solve for } y & : & \tilde{F}^E(x, y) = y - \tilde{F}(x) = 0, \\ \text{Solve for } z & : & z - \bar{f}(x, y) = z - [\bar{F}(y) + x^T x] = 0. \end{array}$$

[5]Note: Typically $\omega(f) >> \sigma(f)$.

The corresponding expanded Hessian matrix H^E is

$$H^E = \begin{pmatrix} -\tilde{J} & I & 0 \\ 0 & \nabla^2_{yy}\bar{f} & I \\ (\tilde{F}^E_{xx})^T w + \nabla^2_{xx}\bar{f} & 0 & -\tilde{J}^T \end{pmatrix},$$

where I is an n-by-n identity matrix and w is equal to $-\nabla_y(\bar{f})$. Thus, a single structured Newton computation for this minimization problem can be written as follows.

1. Initialize the problem size.

```
>> n = 25;     % problem size
>> I = eye(n);
```

2. Initialization. Variable 'tildeFw' represents the inner product, $(\tilde{F}^E)^T w$.

```
>> TildeF = 'broyden';
>> TildeFw = 'tildeFw';
>> BarF = 'barF';
```

3. Initialize x and values in Extra.

```
>> xstart = rand(n,1);
>> Extra.w = ones(n,1);
>> Extra.y = ones(n,1);
>> Extra.n = n;
```

4. Compute the sparsities of the constituent Jacobian and Hessian matrices.

```
>> JPI = getjpi(TildeF, n);
>> hpif = gethpi(BarF, n, Extra);
>> hpiFw = gethpi(TildeFw, n,Extra);
```

5. Compute one step of structured Newton computations.

   (a) Compute y, $\tilde{F}^E_x$, and $\tilde{F}^E_y = I$.

   ```
   >> [y, FEx] = evalj(TildeF, x, [], [], JPI);
   ```

   (b) Compute $\nabla_y \bar{f}$ and $\nabla^2_{yy}\bar{f}$.

   ```
   >> Extra.y = x;
   >> [z, grady, H2]= evalh(BarF, y, Extra, hpif);
   ```

   (c) Set w to $-\nabla_y \bar{f}$.

   ```
   >> Extra.w = -grady(:);
   ```

(d) Compute the function value, gradient, and Hessian of $(\tilde{F}^E)^T w$ with respect to x.

```
>> Extra.y = y;
>> [z, grad, Ht] = evalh(TildeFw, x, Extra, hpiFw);
```

(e) Construct the expanded Hessian matrix H^E.

```
>> HE(1:n, 1: 2*n) = [-FEx, I];
>> HE(n+1:3*n, 2*n+1:3*n) = [I; -FEx'];
>> HE(n+1:2*n, n+1:2*n) = H2;
>> HE(2*n+1:3*n, n+1:2*n) = Ht+2*I;
```

(f) Compute the function value and gradient of the original function.

```
>> myfun = ADfun('OptmFun', 1);
>> [nf , gradx] = feval(myfun, x, Extra);
```

(g) Solve the Newton system and update x.

```
>> HE = sparse(HE);
>> d = -HE  [zeros(2*n,1); gradx(:)];
>> x = x + d(2*n+1:3*n);
```

Note that this section gives only examples of structured Newton computation for solving nonlinear equations and minimization problems. Users can refer to [45] for details about its advantages, applications, performances, and parallelism. In addition, a template structure is provided in Chapter 9.

Chapter 6

Combining C/Fortran with ADMAT

ADMAT can differentiate a differentiable function defined in a MATLAB M-file. ADMAT cannot be applied to any external files, such as MEX files. However, ADMAT can be combined with finite differencing to enable M-file/external file combinations.

Example 6.1. Compute the Jacobian of the Broyden function (which is programmed in C)

See mexbroy.c and CBroy.m

The C file, mexbroy.c, for the Broyden function and its MEX function is as follows:

```
/**************************************************************
%
%       Evaluate the Broyden nonlinear equations test function.
%
%
%     INPUT:
%         x - The current point (row vector).
%
%
%     OUTPUT:
%         y - The (row vector) function value at x.
%
****************************************************************/
#include "mex.h"
#define x_IN 0
#define y_OUT 0

extern void mexbroy(int n, double *x, double *y)
{
    int i;
    y[0] = (3.0-2.0*x[0])*x[0]-2.0*x[1]+1.0;
    y[n-1] = (3.0-2.0*x[n-1])*x[n-1]-x[n-2]+1.0;
```

```
    for(i=1; i<n-1; i++)
        y[i]= (3.0 - 2.0*x[i])*x[i]-x[i-1]-2.0*x[i+1] + 1.0;

}

// MEX Interface function
void mexFunction(int nlhs, mxArray *plhs[], int nrhs, mxArray, *prhs[])
{
    // define input and output variables
    double *x, *y;
    int n;
    if (nrhs != 1)
        mexErrMsgTxt("One input required");
    else if(nlhs >1)
        mexErrMsgTxt("Too many output argument");

    n = mxGetM(prhs[x_IN]);
    if (n == 1)
        n = mxGetN(prhs[x_IN]);

    plhs[y_OUT] = mxCreateDoubleMatrix(n, 1, mxREAL);

    x = mxGetPr(prhs[x_IN]);
    y = mxGetPr(plhs[y_OUT]);

    mexbroy(n, x, y);

    return;
}
```

Note that **mexFunction** is an interface to integrate C/Fortran functions with MATLAB. It transfers input and output arguments between MATLAB and C/Fortran. After defining the Broyden and its MEX function, type "**mex mexbroy.c**" in the MATLAB prompt to compile the MEX file.

Once the compilation succeeds, the Broyden function can be called as a MATLAB function. File **CBroy.m** integrates **mexbroy.c** file into ADMAT via finite differencing.

```
function y = CBroy(x, Extra)
% y = CBroy(x, Extra)
%
%     compute the broyden function at x by C subroutine, mexbroy.c
%     Mapping, CBoy :  R^n ----- > R^n
%
%
%     INPUT:
%         x - The current point (column vector).  When it is an
%             object of deriv class, the fundamental operations
%             will be overloaded automatically.
%         Extra - Parameters required in CBroy.
%
%
%     OUTPUT:
%         y - The (vector) function value at x.  When x is an object
%            of deriv class, y will be an object of deriv class as well.
%             There are two fields in y.  One is the function value
%             at x, the other is the Jacobian matrix at x.
%
%
%     This is an example of how to use the finite difference method to
%     add existing C subroutine into ADMAT package.  Users can call this
%     function as any forward mode functions in ADMAT.
%
%
%                 Cayuga Research
%                     July 2014
%

global globp; global fdeps;

n = length(x);
if isa(x, 'deriv')                   % x is an objective of deriv class
    val = getval(x);           % get the value of x
    drv = getydot(x);        % get the derivative part of x
    y = mexbroy(val);      % compute the function value at x
    ydot = zeros(getval(n), globp);      % initialize the derivative of y
    % compute the derivative by finite difference method
    for i = 1 :  globp
        tmp = mexbroy(val + fdeps*drv(:,i));
        ydot(:,i) = (tmp - y)/fdeps;
    end

    % set y as an objective of deriv class
    y = deriv(y, ydot);
else
    y = mexbroy(x);
end
```

The global variable **fdeps** is the step size used in finite differencing. Its default value is **1e-6**. Users can specify their own step size.

The final finite-difference calculation is illustrated below.

1. Set problem size.

```
>> n = 5;
```

2. Define a **deriv** input variable.

```
>> x = ones(n,1);
>> x = deriv(x, eye(n));
```

3. Compute the Jacobian by finite differencing.

```
>> y = CBroy(x)
val =
      0
     -1
     -1
     -1
      1
deriv =
   -1.0000   -2.0000         0         0         0
   -1.0000   -1.0000   -2.0000         0         0
         0   -1.0000   -1.0000   -2.0000         0
         0         0   -1.0000   -1.0000   -2.0000
         0         0         0   -1.0000   -1.0000
```

4. Extract Jacobian matrix from the finite differencing.

```
>> JFD = getydot(y)
JFD =
   -1.0000   -2.0000         0         0         0
   -1.0000   -1.0000   -2.0000         0         0
         0   -1.0000   -1.0000   -2.0000         0
         0         0   -1.0000   -1.0000   -2.0000
         0         0         0   -1.0000   -1.0000
```

■

Chapter 7

AD for Inverse Problems with an Application to Computational Finance

Many engineering problems, including examples in financial engineering, are expressed as inverse (optimization) problems. Inverse problems, by their definition, are structured; therefore, the ideas in Chapter 5 can be applied to allow for the efficient determination of derivatives, especially first derivatives. Here, we illustrate the use of structured AD (and ADMAT) in solving inverse problems with special attention to the surface volatility problem arising in financial modeling.

7.1 ▪ Nonlinear Inverse Problems and Calibration

The general problem arises when there is a smooth differentiable model of an engineering phenomenon, often manifested as a nonlinear system F. Given the value of a set of parameters $x = \bar{x}$, sometimes called *control* variables, F can be evaluated to yield the corresponding state space y:

$$\text{solve}_y \quad F(x,y) = 0, \tag{7.1}$$

at $x = \bar{x}$. This is called the *forward problem*. For example, in the linear situation the state vector y can be computed, given a rectangular matrix $A \in \mathbb{R}^{m\times n}$ and control variables $x = \bar{x} \in \mathbb{R}^n$, by a simple matrix multiply: $y = A\bar{x}$. In this simple linear case, $F(x,y) \equiv y - A\bar{x}$.

In the linear case, (7.1) reduces to $\quad \text{solve}_y \quad Iy = A\bar{x}$.

The inverse problem flips (7.1) around. Specifically, given a state vector $\bar{y}$, what is the best value of the control variables x such that $||y(x) - \bar{y}||$ is as small as possible, where $y(x)$ is given by (7.1)? Ideally, we would find x such that $y(x) = \bar{y}$, but this is not always possible since typically $\bar{y}$ is an estimate or (7.1) is an approximate model of an ideal situation. In the linear case, $F(x,\bar{y}) = \bar{y} - Ax$, and the inverse problem is

$$\min_x ||y(x) - \bar{y}|| \equiv ||Ax - \bar{y}||. \tag{7.2}$$

When $||\cdot|| = ||\cdot||_2$, (7.2) is the familiar linear least-squares problem. In this chapter, for definiteness, we will henceforth assume[6] that $||\cdot|| = ||\cdot||_2$.

---

[6]We note that the 2-norm is most commonly used, but it is not always the best choice. For example, the 1-norm has the attraction that solutions tend to be sparse (i.e., have many zeros).

This standard (generally, nonlinear) inverse problem is the optimization problem

$$\min_x \|\tilde{F}(x)\|_2 \equiv \min_x \frac{1}{2}\|\tilde{F}(x)\|_2^2, \tag{7.3}$$

where $\tilde{F}(x) \equiv y(x) - \bar{y}$ and $y(x)$ is defined by (7.1). To expose structure in (7.3) as introduced in Chapter 5, we can write the program to evaluate $\tilde{F}(x)$ as

1. Solve for y : $F(x,y) = 0$ (In the linear case $y = Ax$);
2. Output: $z \leftarrow y - \bar{y}$.

To reveal Jacobian structure in line with the view presented in previous chapters, merely think of $\tilde{F}$ as a function of (x,y) and differentiate with respect to both variables to get the extended Jacobian

$$J^E = \begin{bmatrix} F_x & F_y \\ 0 & I \end{bmatrix}.$$

Note that the Jacobian of $\tilde{F}$ with respect to x is given by the Schur complement:

$$J = -F_y^{-1} F_x. \tag{7.4}$$

We make an important observation: While both F_x, F_y may be sparse, the actual Jacobian given by (7.4) is likely dense due to the application of F_y^{-1}. An exception to this is the linear case when $F_y = I$, but generally J will be dense (regardless of how it is computed). Therefore, direct application of AD technology may be an expensive way to get the Jacobian matrix, whereas obtaining J^E can be relatively inexpensive.

7.1.1 ▪ Gauss–Newton

The best algorithm for solving (7.3) depends on several factors, including the relative sizes of m, n as well as the residual size $\|\tilde{F}(x)\|$ at the solution $x = x_*$. When $m > n$ and the residual size at the solution is close to zero, a popular choice of method is a globalized Gauss-Newton approach, e.g., [47]. In this approach a Gauss–Newton step is determined at each iteration, i.e.,

$$\text{solve}_s : \quad Js \simeq -F,$$

in the *least-squares sense* where F, J are evaluated at the current point. Our main point here is that since J is likely dense in an inverse problem (7.4), it may be more efficient to determine J^E using sparse AD applied to $F(x,y)$ and determine the Jacobian matrix J using a sparse factorization of F_y to solve for J in (7.4).

7.1.2 ▪ Example: The Dirichlet Problem

To illustrate an inverse problem, consider the (forward) *linear Dirichlet problem*. Suppose a smooth surface $y(s,t)$ is defined on a closed domain D and satisfies the differential equation

$$y_{ss} + y_{tt} = L(y,s,t)$$

on D, where L is a linear function. Standard discretization, e.g., [82], yields a large sparse positive definite system of equations

$$Ay = b(x), \tag{7.5}$$

where the right-hand-side vector in (7.5) depends on the boundary values, which we denote as x. So the forward problem is defined by (7.5), i.e., given x solve for y. To define an inverse problem, suppose that a "target" surface $\bar{y}$ is given and we wish to determine boundary values x that best yield $\bar{y}$ through (7.5). Following (7.3) our inverse problem is $\min_x \|\tilde{F}(x)\|$, where $\tilde{F} = y(x) - \bar{y}$ and $y(x)$ is defined through (7.5). Following (7.4), we can write

$$J^E = \begin{bmatrix} B_x & A \\ 0 & I \end{bmatrix} \tag{7.6}$$

and, by (7.4),

$$J = -A^{-1}B_x, \tag{7.7}$$

where B_x is the Jacobian of $b(x)$, typically a diagonal matrix. Note that since A, B_x are sparse, J^E is sparse, but Jacobian J is likely dense. Therefore, rather than determining J directly by applying AD to $\tilde{F}$ (or indeed applying the method of finite differences), it is more efficient to apply AD to $\tilde{F}$ with respect to both (x,y) to obtain expression (7.6) and then solve for J using the relationship (7.7).

7.2 ▪ Increased Granularity

The Dirichlet example illustrates the basic structure of most smooth differentiable inverse problems. This example illustrates the efficiency gains when applying AD to this basic structure (as opposed to ignoring the structure altogether). However, many inverse problems come from forward problems with a finer granularity, and it often pays, in efficiency gains, to exploit this finer structure. For example, the forward computation may solve $F(x,y) = 0$, for y, given a setting of the control variables $x = \bar{x}$, where this computation consists of a three-step sequence:

1. Solve for $y_1 : F^1(\bar{x}, y_1) = 0$;
2. Solve for $y_2 : F^2(\bar{x}, y_1, y_2) = 0$;
3. Solve for $y_3 : F^3(\bar{x}, y_1, y_2, y_3) = 0$.

The inverse problem assumes we have a target value for y_3, say $\bar{y}_3$, but the control variables are unknown, so the inverse problem is $\min_x \|\tilde{F}(x)\|$, where $\tilde{F}(x) = y_3(x) - \bar{y}$ and $y_3(x)$ is defined by the forward computation. Specifically, to evaluate $z = \tilde{F}(x)$ we have the following four-step procedure:

1. Solve for $y_1 : F^1(x, y_1) = 0$;
2. Solve for $y_2 : F^2(x, y_1, y_2) = 0$;
3. Solve for $y_3 : F^3(x, y_1, y_2, y_3) = 0$;
4. Output : $z \leftarrow y_3 - \bar{y}_3$.

If we differentiate this procedure with respect to both x, y, then the following extended Jacobian is obtained:

$$J^E = \begin{bmatrix} F_x & F_y \\ 0 & (0,0,I) \end{bmatrix} = \left[\begin{array}{c|ccc} F_x^1 & F_{y_1}^1 & & \\ F_x^2 & F_{y_1}^2 & F_{y_2}^2 & \\ F_x^3 & F_{y_1}^3 & F_{y_2}^3 & F_{y_3}^2 \\ \hline 0 & 0 & 0 & I \end{array}\right]. \tag{7.8}$$

The Jacobian of $\tilde{F}$ with respect to x can be recovered from (7.8):

$$J = -(0,\, 0,\, I)\cdot F_y^{-1} F_x.$$

Again, it is expected that the extended Jacobian J^E will be sparse, and therefore sparse/structured AD techniques can be used to efficiently compute this matrix. The Jacobian matrix J is very likely to be dense due to the application of F_y^{-1}.

7.2.1 ▪ An Example Class Exhibiting Increased Granularity

Time-stepping PDE approaches often yield this type of "high granularity" inverse problem. For example, consider the *heat conductivity problem*. The classical problem involves applying a constant heat source to the left end of a bar (embedded on $[0,1]$ say). Let $z \in [0,1]$ denote the bar and $y(z,t)$ denote the temperature of the bar at time t. We assume that $\frac{\partial y}{\partial t} = 0$ at $z = 0, 1$ for all t. Then the heat equation governs the conductivity of heat:

$$\frac{\partial y}{\partial t} = \frac{\partial}{\partial z}\left(x(z)\frac{\partial y}{\partial z}\right), \tag{7.9}$$

where $x(z)$ is the conductivity (control variables), a material property of the bar. So if $x(z)$ is known, say $x = \bar{x}$, and we have an initial known temperature distribution, $y(z,0) = y_0$, then discretization of (1.30) yields a forward computation. The usual discretization scheme gives

$$\left.\begin{array}{l} \text{Solve for } y^1 \;:\; K(\bar{x})y^0 - y^1 = 0 \\ \text{Solve for } y^2 \;:\; K(\bar{x})y^1 - y^2 = 0 \\ \quad\vdots \\ \text{Solve for } y^m \;:\; K(\bar{x})y^{m-1} - y^m = 0 \end{array}\right\}, \tag{7.10}$$

where y^k denotes the discretized temperature at time t and $K(x)$ is a tridiagonal matrix. If there are m time steps, then the forward computation $F(\bar{x}, y) = 0$ is defined by repeated application of (7.9), where $y = (y^0, y^1, \ldots, y^m)$.

The inverse problem solves for the material property of the bar, x, assuming the final heat distribution, i.e., $y^m = \bar{y}^m$, is known. So the inverse problem is $\min_x \|\tilde{F}(x)\|$, where $z = \tilde{F}(x)$ is defined by

$$\left.\begin{array}{l} \text{Solve for } y^1 \;:\; K(\bar{x})y^0 - y^1 = 0 \\ \text{Solve for } y^2 \;:\; K(\bar{x})y^1 - y^2 = 0 \\ \qquad\vdots \\ \text{Solve for } y^m \;:\; K(\bar{x})y^{m-1} - y^m = 0 \\ \text{Output} : z = y^m - \bar{y}^m \end{array}\right\}. \tag{7.11}$$

The extended Jacobian of $\tilde{F}$, i.e., the Jacobian of $\tilde{F}$ with respect to (x, y), follows the shape of (7.7) but with additional structure since each step depends only on the previous. Specifically,

$$J^E = \begin{pmatrix} K'(x)y^0 & -I & & & & \\ K'(x)y^0 & K(x) & -I & & & \\ K'(x)y^0 & 0 & K(x) & -I & & \\ \vdots & \vdots & \ddots & \ddots & & \\ K'(x)y^0 & 0 & \cdots & 0 & K(x) & -I \\ 0 & 0 & \cdots & 0 & 0 & I \end{pmatrix}.$$

7.2.2 ▪ A General "High-Granularity" Inverse Structure with Smoothing

As illustrated above, inverse problems are driven by optimally choosing the value of control variables to (approximately) satisfy some known information corresponding to a subset of the intermediate variables. To express this generally we define a target state y_T, where y_T is the compressed vector of a designated subset of the components of state vector y. The components of y_T are known. Let $y_S(x)$ be the compressed vector of the components of $y(x)$ corresponding to target components of the inverse problem. So the inverse problem is

$$\min_x \|\tilde{F}(x)\|, \tag{7.12}$$

where $\tilde{F} \equiv y_S(x) - y_T$ and $y(x)$ is defined by (7.1). Hence, the program to evaluate $\tilde{F}(x)$ is simply

1. Solve for y: $F(x,y)=0$;

2. Output: $z \leftarrow y_S - y_T$.

Extending this notation to the general high-granularity case, the program to evaluate $\tilde{F}(x)$ is

$$\left.\begin{array}{l} 1.\ \text{Solve for } y^1\ :\ F^1(x,y_1)=0 \\ 2.\ \text{Solve for } y^2\ :\ F^2(x,y_1,y_2)=0 \\ \quad\vdots \\ m.\ \text{Solve for } y^m\ :\ F^m(x,y_1,y_2,\dots,y_m)=0 \\ m+1.\ \text{Output}:\ z \leftarrow y_S - y_T \end{array}\right\}.$$

In many practical situations, smoothness in the inverse solution x is important. For example, when the inverse solution x is a discretized approximation to a surface or curve, then it is useful/practical to have the computed solution x reflect this. The inverse solution technique developed above does not always generate sufficient smoothness, especially when the number of control variables, x, is large relative to the dimension of the target vector y_T. A standard approach to this situation is to replace problem (7.12) with

$$\min_x \{\|\tilde{F}(x)\|^2 + \lambda \cdot h(x)\}, \tag{7.13}$$

where λ is a positive (weighting) parameter and $h(x)$ is a smoothing function (an example is given below).

7.2.3 ▪ The Implied Volatility Surface Problem for Option Pricing

The implied volatility surface problem is an important inverse problem arising in computational finance. The volatility surface problem is to determine a volatility surface, over price and (future) time, that reflects the expected volatility behavior of an underlying quantity, such as a stock price. The given data are the prices of some recently traded options on the underlying S.

Under various standard assumptions, such as the underlying S follows a one-factor continuous diffusion equation, i.e.,

$$\frac{dS_t}{S_t} = \mu(S_t,t)dt + \sigma(S_t,t)dW_t, \tag{7.14}$$

where $\sigma = \sigma(S_t, t)$ and $\mu = \mu(S_t, t)$ are continuous differentiable functions of the underlying (e.g., stock) S and time t. The volatility surface, $\sigma = \sigma(S_t, t)$ is our main object of interest because it appears in the form of the solution (below), but it is not measurable from the market (since t measures times going forward from the current time).

The value of the European option satisfies the generalized Black–Scholes equation:

$$\frac{\partial y}{\partial t} + \frac{1}{2}\sigma^2 S^2 \frac{\partial^2 y}{\partial S^2} + rS\frac{\partial y}{\partial S} - ry = 0, \tag{7.15}$$

where r is the instantaneous interest rate with known boundary conditions. In (7.15), the sought-for solution y is the fair value of the option, (S, t) are the axes representing the stock price and future time, r is the risk-free interest rate, and $\sigma(S, t)$ is a surface reflecting the volatility of the underlying S. We assume there is no dividend payout. The control variables x can be defined by discretizing the volatility surface $\sigma(S, t)$. For example, an $N \times N$ discretization across the $S \times t$ plane yields a vector x with $n = N^2$ components. The other parameters in (21) are explicitly determined, so if the volatility surface is also known, $x = \bar{x}$, and boundary values are known, then a standard PDE technique can be used to define a forward computation to (approximately) solve (7.16), as in (7.1):

$$\text{Solve}_y \quad F(x,y) = 0 \quad \text{at} \quad x = \bar{x}. \tag{7.16}$$

Some PDE techniques for this problem, such as Crank–Nicholson, are time-stepping techniques, and so the high-granularity structure, similar to (7.10), (7.11), can be applied to break (7.16) into a series of substeps. The general situation here is that there are m financial options to evaluate all defined on the same underlying S and all satisfying (7.13) but with different boundary conditions. So in this case $y = y(x)$ is a vector of length m, and component y_i is determined by the forward computation (7.16) using the appropriate boundary conditions. Hence, the forward computation of vector y involves m PDE solutions (7.16).

The inverse problem starts with m known option prices, $\bar{y}^T = (\bar{y}_1, \ldots, \bar{y}_m)$. All options are on the same underlying S, all traded recently at about the same time to yield $\bar{y}$, and all assumed to satisfy (7.15).

The inverse problem is to determine a discrete approximation to the volatility surface x is just (7.3), where component $y_i(x)$ is computed by solving $F(x, y_i) = 0$ for y_i corresponding to the PDE evaluation of option i. The high-level structure of the inverse problem is the usual:

$$\min_x \|\tilde{F}(x)\| = \min_x \frac{1}{2}\|\tilde{F}(x)\|_2^2,$$

where $\tilde{F}(x) \equiv y(x) - \bar{y}$ and $y(x)$ is defined by (7.16). We can write the program to evaluate $\tilde{F}(x)$ as

1. Solve for $y: \; F(x,y) = 0$;
2. Output: $z \leftarrow y - \bar{y}$.

This inverse problem is typically significantly underdetermined, usually $m \ll n$. For example, a 100×100 discretization yields $n = 10{,}000$, whereas in practice it is not uncommon to have $m \approx 20$. Therefore, there are typically many solutions x (discretized volatility surfaces) that satisfy the data, but the corresponding surfaces can be very rough (i.e., nonsmooth): often the solution x reflects an unrealistic and unusable solution. Therefore, it was suggested in [73] to add a smoothing term to the inverse problem, e.g., (7.13). For example, in [73] it is suggested to define $h(x)$ such that

$$h(x) = ||\nabla\sigma(s,t)||_2$$

by finite differencing at the $n = N^2$ grid points. Choosing the "best" λ is not an easy problem: this choice balances smoothness of the volatility surface with satisfaction of the data points.

7.2.4 ▪ Variable Reduction and Splines

A challenge of the approach described above is the size of the nonlinear minimization problem to be solved, i.e., $n = N^2$, where N reflects the fineness of the discretization used for the volatility surface computation. Coleman et al. [33] propose a method to reduce the dimensionality by introducing a cubic spline representation of the volatility surface with $p \approx m$ knot points. In the end the inverse problem in this new setting involves p unknowns (the value of the spline approximations at the knot points) as opposed to $n = N^2$ unknowns.

The idea developed in [33] is based on the observations that typically only a few recent option prices on the underlying S are known (e.g., $m \approx 20$), and a smooth bi-cubic spline surface can be defined on a few knots points, say $p \approx m$ and known boundary values. Moreover, the volatility surface to be approximated is known, under reasonable assumptions, to be smooth, and bi-cubic surfaces have strong smoothness properties.

In this setting the control variables become the values $x_1,\ldots,x_p$ of the bi-cubic spline at the p knot points. The spline surface itself, discretely approximated by $n = N^2$ grid points, is uniquely defined by a smooth differentiable mapping: $y_1 = F^1(x)$, given boundary conditions. So the forward problem can be written at $x = \bar{x}$:

$$\left.\begin{array}{l} \text{1. Solve for } y_1: \; F^1(\bar{x}, y_1) = 0 \\ \text{2. Solve for } y_2: \; F^2(y_1, y_2) = 0 \end{array}\right\}, \tag{7.17}$$

where y_2 is the m-vector of option values. Note that if F^2 is a time-stepping PDE evaluation, then this second step can be broken down into T substeps, where T is the number of time steps similar to (7.10), (7.11). Following the forward problem (7.9) the inverse problem is $\min_x ||\tilde{F}(x)||$, where $\tilde{F}(x) \equiv y_2(x) - \bar{y}_2$ and $\bar{y}_2$ is the m-vector corresponding to the known option prices. So to evaluate $z = \tilde{F}(x)$ we have the following three-step procedure:

$$\left.\begin{array}{l} \text{1. Solve for } y_1: \; F^1(\bar{x}, y_1) = 0 \\ \text{2. Solve for } y_2: \; F^2(y_1, y_2) = 0 \\ \text{3. Output}: \; z \leftarrow y_2 - \bar{y}_2 \end{array}\right\}.$$

The corresponding extended Jacobian is

$$J^E = \begin{bmatrix} F_x^1 & -I & 0 \\ 0 & F_{y_1}^2 & F_{y_2}^2 \\ 0 & 0 & I \end{bmatrix},$$

where F_x^1 is $n \times p$, $F_{y_1}^2$ is $m \times n$, and $F_{y_2}^2$ is $m \times m$. Note that a Schur-complement computation yields the Jacobian of $\tilde{F}$ (with respect to x):

$$J = (F_{y_2}^2)^{-1} F_{y_1}^2 F_x^1,$$

which is a compact matrix of size $m \times p$.

Conveniently, the calculation of the $m \times p$ matrix J is (usually) most efficiently determined by applying forward-mode AD to $\tilde{F}$: $\omega \sim p \times \omega(F)$, $\sigma \sim \sigma(F)$.

Chapter 8

A Template for Structured Problems

The companion AD software package, ADMAT, allows the user to compute first and second derivatives in the MATLAB environment. Chapter 4 (and parts of 7) illustrate how to efficiently apply ADMAT for structured problems. This is quite important since many practical problems are structured, and very significant efficiency gains—in space and time—are obtained by taking advantage of structure. Indeed, the structured use of ADMAT, efficiently balancing the use of forward and reverse modes of AD, can turn otherwise intractable problems of differentiation into efficient procedures. The efficiency gains are due, in part, to discovering and exploiting hidden sparsity in an apparently dense computing problem.

In this chapter we introduce ADMAT-ST, a collection of templates to help the user implement the structured application of AD as discussed in Chapter 4. ADMAT-ST has templates for structured Jacobian, gradient, and Hessian computations.

8.1 ▪ The Jacobian Template

We assume that the differentiable vector-valued function, $z = F(x)$, where $F : \mathbf{R}^m \to \mathbf{R}^n$, can be written in the structured form given in (4.5) and reproduced below:

$$\left.\begin{array}{rcl} \text{Solve for } y_1 & : & F_1^E(x,y_1)=0 \\ \text{Solve for } y_2 & : & F_2^E(x,y_1,y_2)=0 \\ \vdots & & \vdots \\ \text{Solve for } y_p & : & F_p^E(x,y_1,y_2,\ldots,y_p)=0 \\ \text{Solve for output } z & : & z-F_{p+1}^E(x,y_1,y_2,\ldots,y_p)=0 \end{array}\right\}. \tag{8.1}$$

The ADMAT-ST function JACOBIANST efficiently computes the $m \times n$ Jacobian of F using the structure ideas developed in Chapter 4, provided the structure is identified through input parameters given to JACOBIANST:

```
function [z,M] = jacobianst(functions, x, dependency, Extra)
% requires ADMAT methods:  ADfun, feval
% - - - - - - - - - - - - - - - - - - - - - - - - - -
% jacobianst computes the first derivatives of a structured
%     computation
% z - output of the computation
% M - output of first derivative information in the form W^T *J * V
%     where J is the Jacobian and W and V are optional matrices
%     specified in Extra by default W and V are identities
% functions - cell array of function names of each step of the
%     computation
% x - vector of initial input
% dependency - N-by-2 cell array of dependecy information of the
%     steps of the computation
% Extra - optional structure containing any user defined parameter
%     required by the computation
```

Note that the output matrix M is the Jacobian matrix J or the product JV or the product W^TJ, depending on the input parameters. Below is a simple example of computing the Jacobian matrix J of a simple structured problem. Consider a computation of the following form:

```
function z = F(x)
y1 = f1(x);
y2 = f2(x);
y3 = f3(y1,y2);
y4 = f4(y1,y2);
y5 = f5(y3,y4);
z = f6(y5);
```

for which the Jacobian can be computed using JACOBIANST:

```
functions = {'f1'; 'f2'; 'f3'; 'f4'; 'f5'; 'f6'};
dependency = cell(6,2);
dependency(1,:)  = {1,[]};     % f1 and f2 depend only on x
dependency(2,:)  = {1,[]};
dependency(3,:)  = {0,[1,2]};     % f3 and f3 depend on y1 and y2
dependency(4,:)  = {0,[1,2]};
dependency(5,:)  = {0,[3,4]};     % f5 depends on y3 and y4
dependency(6,:)  = {0,5};     % f6 depends on y5
[ z, J ] = jacobianst(functions, x, dependency, [ ])
```

In the above example, neither W nor V is specified, so both are assumed to be identity, and the full Jacobian matrix is returned.

8.2 ▪ The Gradient Template

The gradient of a scalar-valued function is a special case of a Jacobian computation. A structured gradient can be computed using the tools described in Section 1.1. However, due to the importance and popularity of gradient computations, ADMAT-ST has specialized functions for structured gradient computations. We assume the scalar-valued function, $z = f(x)$, can be written in the structured form given in (4.16). The ADMAT-ST function GRADIENTST efficiently computes the gradient using the structure ideas developed in Chapter 4, provided the structure is identified through input parameters given to GRADIENTST:

```
function [z,grad] = gradientst(functions, x, dependency, Extra)
% requires ADMAT methods:  jacobianst
% - - - - - - - - - - - - - - - - - - - - - - - - -
% gradientst computes the gradient of the computation
% z - output of the computation
% grad - the gradient
% functions - cell array of function names of each step of the
%     computation the last name should correspond to a scalar function
% x - vector of initial input
% dependency - N-by-2 cell array of dependecy information of the steps
%     of the computation
% Extra - optional structure containing any user defined parameter
%     required by the computation
```

To illustrate with simple examples, consider Broyden iteration with initial point x defined as

$$
\begin{aligned}
y^{(0)} &= x,\\
y_1^{(k)} &= (3-2y_1^{(k-1)})y_1^{(k-1)} - 2y_2^{(k-1)} + 1,\\
y_i^{(k)} &= (3-2y_i^{(k-1)})y_i^{(k-1)} - 2y_{i-1}^{(k-1)} + 1, \quad i = 2,\dots,n-1\\
y_n^{(k)} &= (3-2y_n^{(k-1)})y_n^{(k-1)} - 2y_{n-1}^{(k-1)} + 1
\end{aligned}
$$

for iteration $k = 1,2,\dots$. The gradient of the squared norm $y^T y$ of the final solution with respect to the initial point can be computed using GRADIENTST:

```
functions(1:MaxIter) = {'broyden'}; % apply broyden at each iteration
functions(MaxIter+1) = {'squaredNorm'}; % compute squared norm at last step
% first iteration takes in initial input x
dependency(1,:) = {1,[]};
% iterations afterwards doesn't use x
dependency(2:(MaxIter+1),1) = {0};
% each iteration only takes in the output from its precursor
dependency(2:(MaxIter+1),2) = num2cell(1:MaxIter)';
[z,grad] = gradientst(functions,x,dependency,[]);
```

where a function called "broyden" implements Broyden iteration defined above and MaxIter is the number of iterations required by the user.

8.3 ▪ The Hessian Template

ADMAT-ST has template functions to allow for the efficient computation of the matrix of n derivatives, the Hessian matrix, of a (structured) twice continuously differentiable scalar-valued function, $z = f(x)$. The structure is given in (4.28), where the Hessian is provided in (4.35).

The ADMAT-ST function HESSIANST efficiently computes the $n \times n$ Hessian of f using the structure ideas developed in Chapter 4, provided the structure is identified through input parameters given to HESSIANST:

```
function [z, grad, H, Extra] = hessianst(functions, x, dependency, Extra)
% requires ADMAT methods:  feval, evalj, evalh
% - - - - - - - - - - - - - - - - - - - - - - - - - -
% hessianst computes the second derivatives of a structured computation
% z - output of the computation
% grad - gradient of the computation
% H - Hessian of the computation
% functions - cell array of function names of each step of the computation
%      the last name should correspond to a scalar function
% x - vector of initial input
% dependency - N-by-2 cell array of dependency information of the steps of
%      the computation
% Extra - optional structure containing sparsity information JPIs, HPIs,
%      and any user defined parameter required by the computation
%
% If sparsity information is not provided, the function will computed and
%      return it as Extra.JPIs.
```

The user can provide the sparsity pattern of each function by specifying Extra.JPIs for Jacobian and Extra.HPIs for Hessian. If sparsity patterns are not provided, HESSIANST will automatically calculate the sparsity pattern and return it via the struct Extra.

To illustrate, below is a simple example. Consider a computation of the following form:

```
function z = F(x)
y1 = f1(x);
y2 = f2(x);
y3 = f3(y1,y2);
y4 = f4(y1,y2);
y5 = f5(y3,y4);
z = f6(y5);
```

for which the Hessian can be computed using HESSIANST:

```
functions = {'f1'; 'f2'; 'f3'; 'f4'; 'f5'; 'f6'};
dependency = cell(6,2);
dependency(1,:)  = {1,[]};     % f1 and f2 only depend on x
dependency(2,:)  = {1,[]};
dependency(3,:)  = {0,[1,2]};     % f3 and f3 depend on y1 and y2
dependency(4,:)  = {0,[1,2]};
dependency(5,:)  = {0,[3,4]};     % f5 depends on y3 and y4
dependency(6,:)  = {0,5};    % f6 depends on y5
[z, grad, H, Extra] = hessianst(functions, x, dependency, Extra)
```

Chapter 9

R&D Directions

Automatic differentiation (AD) is a technology whose time has come. AD can efficiently and accurately determine derivatives—especially first and second derivatives—for use in multi-dimensional minimization and nonlinear systems solutions. ADMAT is an AD tool designed for use in the flexible MATLAB computing environment. In addition to providing basic differentiation services (e.g., forward and reverse modes of differentiation), ADMAT can be effectively used for sparse and/or structured application problems.

It is the reverse mode of AD that clearly separates AD from other possible differentiation approaches (such as finite-differencing and handcoding). The extensive memory requirements of reverse mode can be greatly mitigated by using the structure ideas highlighted in this book. With these structure concepts in hand, reverse mode becomes a powerful and practical mode of differentiation. In addition, combining forward mode with reverse mode can further increase the range (and efficiency) of AD, especially when exploiting structure (see Chapter 3).

There remains research and development work to be done on AD. On the development side, focusing specifically on ADMAT, work must be done to extend applicability to 3D data structures. Currently ADMAT 2.0 can handle 3D structures only as a collection of 2D structure—this restricts the applications user from some MATLAB coding techniques that might otherwise be appropriate.

Another promising R&D direction is to integrate AD technology, specifically ADMAT, with graphics processor unit (GPU) computing. A GPU is a specialized electronic circuit designed to rapidly manipulate and alter memory to accelerate the creation of images in a frame buffer intended for output to a display. Most computers are equipped with a GPU that handles their graphical output, including the 3D animated graphics used in computer games. The computing power of GPUs has increased rapidly, and they are now often much faster than the computer's main processor, or CPU.

Recently, GPU technology is increasingly used in scientific computing to speed up expensive matrix computations. Research and development is needed to connect AD with GPU technology in the MATLAB environment.

For large problems, parallelism can be used in the derivative computation through AD.

One promising parallelism approach for AD is to use GPU technology.

For forward-mode AD, an operation overload class can be defined on the GPU end so that the derivative computations can be distributed on GPU cores evenly. Then, when all these independent computations on GPU cores are completed, all these computed results can be collected on the CPU end as outputs.

For reverse-mode AD, a stack can be defined on the CPU end, which works as a "tape" storing all the intermediate information for the function evaluation. Then the cores on the GPU end will access this stack to evaluate each component in the derivatives simultaneously. Finally, all computed entries are transferred to the CPU end for return. Due to the definition of Jacobian, gradient, and Hessian matrix and the "hidden" parallelism in these derivatives, the GPU implementation is suitable for large applications whose derivatives need to be evaluated many times with different input variables, such as Newton-like methods.

We conclude with a final broad research direction. Note that while we expect our template framework presented in Chapter 8 to be very useful when developing new applications, it is clearly much less useful when faced with legacy codes and solutions. Moreover, we realize that even with an easy-to-use template framework, some applications developers will resist this direction, preferring to write code in a less-than-compatible fashion. Therefore, we remain very interested in developing methods that will automatically recognize AD-friendly structure (of the sort highlighted in this book) given a user source code written in any (correct) fashion. A possible start in this direction is given in [43].

For example, if a function $F : \mathscr{R}^2 \to \mathscr{R}^3$ is defined as

$$F\left(\begin{bmatrix} x_1 \\ x_2 \end{bmatrix}\right) = \begin{bmatrix} \sin(\cos(\sin 2^{x_1} + x_2^2) \cdot (5x_1 - 6x_2)) \\ (2x_1^{x_2} + x_2^{x_1})^{\sin x_1 + \cos x_2} \\ \cos(\sin 2^{x_1} + x_2^2) + (5x_1 - 6x_2) + (2x_1^{x_2} + x_2^{x_1}) + (\sin x_1 + \cos x_2 \end{bmatrix},$$

then F's computational graph $\bar{G} = \{V, \bar{E}\}$, where V is the set of atomic computations and $\bar{E}$ is the set of directed edges reflecting the flow of intermediate variables, as indicated in Figure 9.1. The basic idea is one of recursively finding small directed edge separators that divide the remaining graph into approximately equal pieces. It turns out that these separators can be identified as the vectors of distinguished intermediate variables "y" used throughout Chapter 5.

Definition 9.1. *$E_{cut} \in \bar{E}$ is a directed edge separator in directed graph $\bar{G}$ if $\bar{G} - \{E_{cut}\} = \{G_1, G_2\}$, where G_1 and G_2 are disjoint and all edges in E_{cut} have the same orientation relative to G_1 and G_2.*

So step 1 of this recursive approach is to find a small (directed) edge separator that divides the graph in Figure 9.1 roughly evenly. One choice is indicated in Figure 9.2. Suppose $E_{cut} \subset \bar{E}_y$ is a directed separator of the computational graph $\bar{G}(F)$ with orientation forward in time. Then the nonlinear function $F(x)$ can be broken into two parts:

$$\left.\begin{array}{l} \text{Solve for } y: \ F_1(x,y) = 0 \\ \text{Solve for } z: \ F_2(x,y) - z = 0 \end{array}\right\}, \qquad (9.1)$$

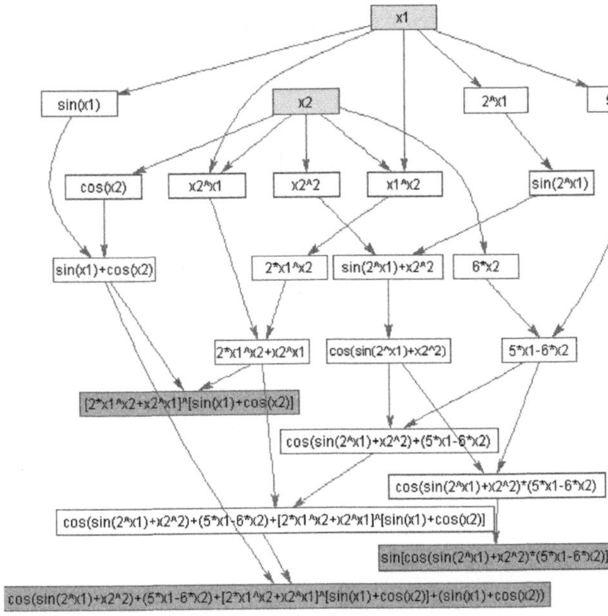

Figure 9.1. *An example computational graph.*

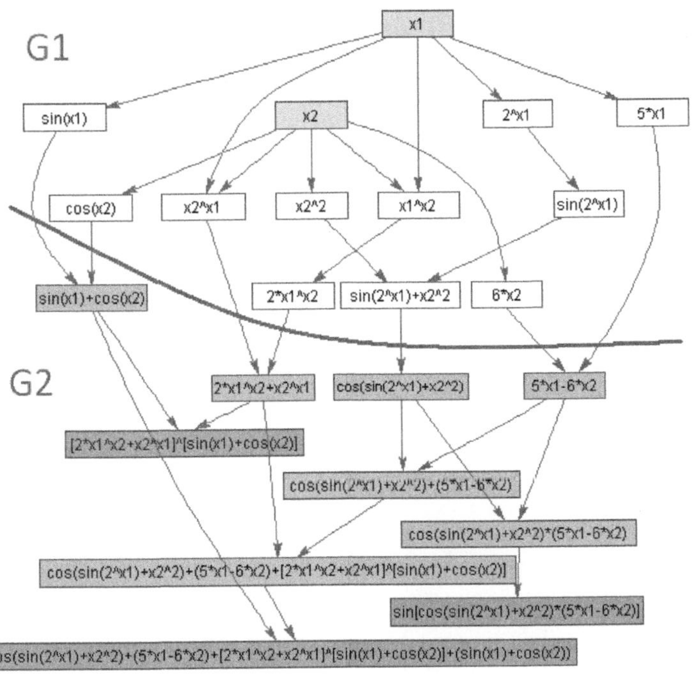

Figure 9.2. *A direct separator.*

where y is the vector of intermediate variables defined by the tail vertices of the edge cut set E_{cut} and z is the output vector, i.e., $z = F(x)$. Let p be the number of tail vertices of edge set E_{cut}, i.e., $y \in \mathcal{R}^p$. Note that $|E_{cut}| \geq p$. The nonlinear function F_1 is defined by the computational graph to the above of E_{cut}, i.e., G_1, and nonlinear function F_2 is defined by the computational graph below E_{cut}, i.e., G_2. See Figure 9.2. Following the approach developed in Chapter 5, we note that the system (1.2) can be differentiated with respective to (x, y) to yield an "extended" Jacobian matrix:

$$J_E = \begin{bmatrix} (F_1)_x & (F_1)_y \\ (F_2)_x & (F_2)_y \end{bmatrix}. \tag{9.2}$$

Since y is a well-defined unique output of function $F_1 : \mathcal{R}^n \rightarrow \mathcal{R}^{n+p}$, $(F_1)_y$ is a $p \times p$ nonsingular matrix. The Jacobian of F is the Schur complement of (9.2), i.e.,

$$J(x) = (F_2)_x - (F_2)_y (F_1)_y^{-1} (F_1)_x.$$

In principle this separator idea can be applied recursively to the separated graph pieces.

Research issues include how to effectively find a good set of separators given the computational graph and how to find good separators when examining only a "current" section of the graph (since for realistic problems it is not feasible to manipulate the entire computational graph).

Appendix A

Installation of ADMAT

In this appendix, the installation of ADMAT on a Unix (Linux) and Windows platform is discussed.

A.1 ▪ Requirements

ADMAT belongs to the "operator overloading" class of AD tools and uses object-oriented programming features. Thus, ADMAT requires MATLAB 6.5 or above.

A.2 ▪ Obtaining ADMAT

The complete ADMAT package, *ADMAT* 2.0 *Professional*, can be obtained from Cayuga Research Inc. See www.cayugaresearch.com for details.

A.3 ▪ Installation Instructions for Windows Users

ADMAT is supplied in the zipped file ADMAT-2.0.zip. Follow these installation instructions.

1. Place the ADMAT package in an appropriate directory and unzip the package using any unzip software.

2. There are two ways to set the search path in MATLAB.

   ****************** Method 1. ******************

   (a) Click "File" in MATLAB window.
   (b) Choose "Set Path" option.
   (c) Click "Add with Subfolders" button.
   (d) Find the targeted directory for the ADMAT package in the "Browse for Folders" window and click "OK."
   (e) Click the "Save" button to save the path for ADMAT and click the "Close" button.
   (f) Type "startup" in the MATLAB prompt or log out of MATLAB and log in again.

******************* Method 2. *******************

Access the ADMAT directory, edit the **startup.m** file, add ALL subdirectories of ADMAT, search paths in the file manually, save the file, and type “startup” in the MATLAB prompt to set up the paths for the package.

3. There is a “success” message as ADMAT is correctly installed. Users can type “help ADMAT-2.0” in the MATLAB prompt to get a list of main functions in the package.

A.4 ▪ Installation Instructions for Unix (Linux) Users

1. Unzip ADMAT-2.0.zip using unzip ADMAT-2.0.zip in the Unix (Linux) prompt.
2. Follow Method 2 listed above.

Appendix B

How Are Codes Differentiated?

AD is a chain rule–based technique for calculating derivatives with respect to the input variables of functions programmed in a high-level computer language, such as C and MATLAB. In AD, it is assumed that all mathematical functions, no matter how complicated, are defined by a finite set of elementary operations (such as unitary or binary operations, e.g., $\sqrt{(\cdot)}$, $\cos(\cdot)$). The function value computed by the corresponding program is simply a composition of these elementary functions. The partial derivatives of the elementary functions usually are given in the AD package; then the derivative for the specified function can be computed via the chain rule.

Let $z = f(x)$, where $f : \mathcal{R}^n \to \mathcal{R}^m$. The function value, z, at some given point x, can be computed in the following form:

$$\begin{aligned} x &\equiv (x_1, x_2, \ldots, x_n) \\ &\downarrow \\ y &\equiv (y_1, y_2, \ldots, y_p) \\ &\downarrow \\ z &\equiv (z_1, z_2, \ldots, z_m), \end{aligned} \tag{B.1}$$

where the intermediate variables y are constructed through a series of elementary functions. For a unitary elementary function (such as $\sqrt{(\cdot)}$, $\sin(\cdot)$), we have

$$y_k = f_{elem}^k(y_i), \quad i < k.$$

For a binary elementary function (such as $+, -, /$), we have

$$y_k = f_{elem}^k(y_i, y_j), \quad i, j < k.$$

In general, the number of intermediate variables is much larger than the dimension of the problem, i.e., $p \gg m, n$.

There are two basic modes for implementing AD: the forward mode and the reverse mode. In the forward mode, the derivatives are propagated throughout the entire computation, using the chain rule. For example, for the elementary function $y_k = f_{elem}^k(y_i, y_j)$, the corresponding derivative for the intermediate variables y_k, $\frac{dy_k}{dx}$, can be computed in the forward mode as

$$\frac{dy_k}{dx} = \frac{\partial f_{elem}^k}{\partial y_i} \cdot \frac{dy_i}{dx} + \frac{\partial f_{elem}^k}{\partial y_j} \cdot \frac{dy_j}{dx}.$$

The chain rule is applied to all intermediate variables y_k, $k = 1, \ldots, p$, until the output variable z and $\frac{dz}{dx}$ are computed.

The reverse mode computes the derivatives $\frac{dz}{dy_k}$ for all intermediate variables backward. For example, for the elementary step $y_k = f_{elem}^k(y_i, y_j)$, the corresponding derivatives $\frac{dz}{dy_i}$ and $\frac{dz}{dy_j}$ can be computed in the reverse mode as

$$\frac{dz}{dy_i} = \frac{dz}{dy_i} + \frac{\partial f_{elem}^k}{\partial y_i} \frac{dz}{dy_k} \quad \text{and} \quad \frac{dz}{dy_j} = \frac{dz}{dy_j} + \frac{\partial f_{elem}^k}{\partial y_j} \frac{dz}{dy_k}, \tag{B.2}$$

where the initial values for $\frac{dz}{dy_i}$ and $\frac{dz}{dy_j}$ are zero. The whole process starts at $\frac{dz}{dz} = I_{m\times m}$, where $I_{m\times m}$ is an $m \times m$ identity matrix, and ends at $\frac{dz}{dx}$. The reverse mode has to save the entire computational trace, so-called tape, since the propagation is done backward through the computation. Hence, all partials $\frac{dz}{dy_i}$ and $\frac{dz}{dy_j}$ have to be stored for derivation computation as in (B.2). It implies that the reverse mode can be prohibitive for some complicated functions due to memory requirements.

In summary, given a function $F(x) : \mathcal{R}^n \to \mathcal{R}^m$, the forward-mode AD computes the Jacobian matrix product $J \cdot V$, while the reverse-mode AD computes the adjoint product $J^T \cdot W$, where J is the first derivative of $F(x)$, $V \in \mathcal{R}^{n\times p_1}$, and $W \in \mathcal{R}^{m\times p_2}$.

Next, we will show how to obtain the Jacobian matrix products of $J \cdot V$ and $J^T \cdot W$ for the function $F(x)$ defined in Example 2.1 via the forward and reverse modes with $V = W = \left[\begin{smallmatrix} 1 & 0 \\ 0 & 1 \end{smallmatrix}\right]$, respectively.

Example B.1. Compute the Jacobian matrix product $J \cdot V$ of $F(x) = \begin{pmatrix} 3x_1+2x_2 \\ 5x_1^2+4x_1+x_2 \end{pmatrix}$ at point $\bar{x} = \left(\begin{smallmatrix} 1 \\ 2 \end{smallmatrix}\right)$ with $V = \left[\begin{smallmatrix} 1 & 0 \\ 0 & 1 \end{smallmatrix}\right]$ via the forward mode. ■

Based on the process in (B.1), the intermediate variable $y \equiv (y_1, \ldots, y_6)$ can be defined as

$$y_1 = 3x_1, \ y_2 = 2x_2, \ y_3 = x_1^2, \ y_4 = 5y_3, \ y_5 = 4x_1, \ y_6 = y_5 + x_2,$$

and finally the output is obtained:

$$z_1 = y_1 + y_2, \qquad z_2 = y_4 + y_6.$$

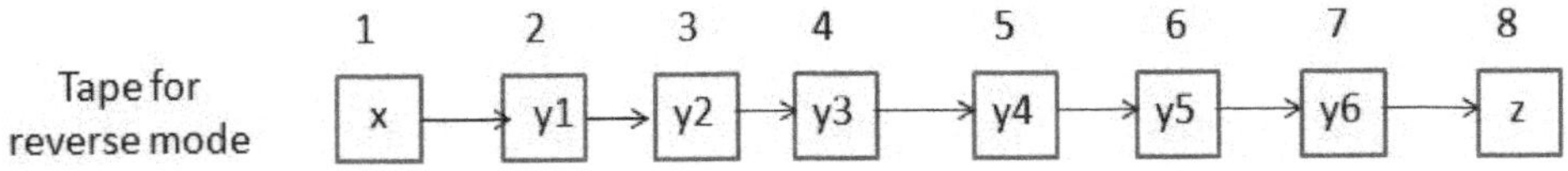

Figure B.1. *Tape for computing $J^T \cdot W$ via the reverse mode in Example B.2.*

Then the computation process to compute $J \cdot V$ is as follows:

$$
\begin{aligned}
&\frac{dx}{dx} = V = \begin{bmatrix} 1 & 0 \\ 0 & 1 \end{bmatrix}, \\
&\frac{dy_1}{dx_1} = 3 * \frac{dx_1}{dx_1} = 3, \quad \frac{dy_1}{dx_2} = 0, \\
&\frac{dy_2}{dx_1} = 0, \quad \frac{dy_2}{dx_2} = 2 * \frac{dx_2}{dx_2} = 2, \\
&\frac{dy_3}{dx_1} = 2 * x_1 \frac{dx_1}{dx_1} = 2, \quad \frac{dy_3}{dx_2} = 0, \\
&\frac{dy_4}{dx_1} = 5 * \frac{dy_3}{dx_1} = 10, \quad \frac{dy_4}{dx_2} = 0, \\
&\frac{dy_5}{dx_1} = 4 * \frac{dx_1}{dx_1} = 4, \quad \frac{dy_5}{dx_2} = 0, \\
&\frac{dy_6}{dx_1} = \frac{dy_5}{dx_1} = 4, \quad \frac{dy_6}{dx_2} = \frac{dx_2}{dx_2} = 1, \\
&\frac{dz_1}{dx_1} = \frac{dy_1}{dx_1} + \frac{dy_2}{dx_1} = 3, \quad \frac{dz_1}{dx_2} = \frac{dy_1}{dx_2} + \frac{dy_2}{dx_2} = 2, \\
&\frac{dz_2}{dx_1} = \frac{dy_4}{dx_1} + \frac{dy_6}{dx_1} = 14, \quad \frac{dz_2}{dx_2} = \frac{dy_4}{dx_2} + \frac{dy_6}{dx_2} = 1.
\end{aligned}
$$

Thus, the Jacobian matrix product $J \cdot V = \left[\begin{smallmatrix} 3 & 2 \\ 14 & 1 \end{smallmatrix}\right]$.

Example B.2. Compute the Jacobian matrix product $J^T \cdot W$ of $F(x)$ at point $\bar{x}$ as in Example B.1 with $W = \left[\begin{smallmatrix} 1 & 0 \\ 0 & 1 \end{smallmatrix}\right]$ via the reverse mode. ■

The intermediate variables are the same as in Example B.1. Note that the initial values for intermediate variables and input variables are zero, i.e., $\frac{dz_i}{dy_j} = 0$, $i = 1, 2, \ldots, m$, $j = 1, \ldots, p$. We first initialize the derivatives for z as $\frac{dz}{dz} = W = \left[\begin{smallmatrix} 1 & 0 \\ 0 & 1 \end{smallmatrix}\right]$.

Figure B.1 illustrates the "tape" to store all computational traces of input variable, x; intermediate variables, $y_1, y_2, \ldots, y_6$; and output variable z. For example, the seventh cell in the tape stores the following computational information for intermediate variable y_6, i.e., $y_6 = y_5 + x_2$, as

```
op: 'plus'
val: 12
arg1: 5
arg2: 1
W: [0, 1]
```

The description for all fields in this cell is as follows:

op: elementary operation for y_6
val: value of y_6
arg1: cell number of the first input argument of the elementary operation in 'op' field.
arg2: cell number of the second input argument of the elementary operation in 'op' field.
W: derivative values of $\frac{dz}{dy_6}$.

Then the derivative computing process starts from cell 8 and rolls back to cell 1 at the beginning of the tape. The roll-back process is as follows:

$$\frac{dz_1}{dy_1} = \frac{dz_1}{dy_1} + 1 * \frac{dz_1}{dz_1} = 1, \quad \frac{dz_1}{dy_2} = \frac{dz_1}{dy_2} + 1 * \frac{dz_2}{dz_2} = 1$$
$$\frac{dz_2}{dy_4} = \frac{dz_2}{dy_4} + 1 * \frac{dz_2}{dz_2} = 1, \quad \frac{dz_2}{dy_6} = \frac{dz_2}{dy_6} + 1 * \frac{dz_2}{dz_2} = 1$$
$$\frac{dz_1}{dy_5} = \frac{dz_1}{dy_5} + 1 * \frac{dz_1}{dy_6}, \quad \frac{dz_1}{dx_2} = \frac{dz_1}{dx_2} + 1 * \frac{dz_1}{dy_6} = 0$$
$$\frac{dz_2}{dy_5} = \frac{dz_2}{dy_5} + 1 * \frac{dz_2}{dy_6} = 1, \quad \frac{dz_2}{dx_2} = \frac{dz_2}{dx_2} + 1 * \frac{dz_2}{dy_6} = 1,$$
$$\frac{dz_1}{dx_1} = \frac{dz_1}{dx_1} + 4 * \frac{dz_1}{dy_5} = 0, \quad \frac{dz_2}{dx_1} = \frac{dz_2}{dx_1} + 4 * \frac{dz_2}{dy_5} = 4,$$
$$\frac{dz_1}{dy_3} = \frac{dz_1}{dy_3} + 5 * \frac{dz_1}{dy_4} = 0, \quad \frac{dz_2}{dy_3} = \frac{dz_2}{dy_3} + 5 * \frac{dz_2}{dy_4} = 5,$$
$$\frac{dz_1}{dx_1} = \frac{dz_2}{dx_1} + 2 * x_1 \frac{dz_1}{dy_3} = 0, \quad \frac{dz_2}{dx_1} = \frac{dz_2}{dx_1} + 2 * x_1 * \frac{dz_2}{dy_3} = 14,$$
$$\frac{dz_1}{dx_2} = \frac{dz_1}{dz_2} + 2 * \frac{dz_1}{dy_2} = 2, \quad \frac{dz_2}{dx_2} = \frac{dz_2}{dx_2} + 2 * \frac{dz_2}{dy_2} = 1,$$
$$\frac{dz_1}{dx_1} = \frac{dz_1}{dx_1} + 3 * \frac{dz_1}{dy_1} = 3, \quad \frac{dz_2}{dx_1} = \frac{dz_2}{dx_1} + 3 * \frac{dz_2}{dy_1} = 14.$$

Thus, the adjoint product is $J^T \cdot W = \left[\begin{smallmatrix} 3 & 14 \\ 2 & 1 \end{smallmatrix}\right]$.

Bibliography

[1] ADOL-C package, https://projects.coin-or.org/ADOL-C.

[2] L. B. G. Andersen and R. Brotherton-Ratcliffe, *The equity option volatility smile: An implicit finite difference approach*, Journal of Computational Finance, Vol. 1, 1997, 5–37.

[3] Autodiff.org, www.autodiff.org, 2015.

[4] M. Bartholomew-Biggs, S. Brown, B. Christianson, and L. Dixon, *Automatic differentiation of algorithms*, Journal of Computational and Applied Mathematics, Vol. 124, 2000, 171–190.

[5] W. Baur and V. Strassen, *The complexity of partial derivatives*, Theoretical Computer Science, Vol. 22, 1983, 317–330.

[6] L. M. Beda, L. N. Korolev, N. V. Sukkikh, and T. S. Frolova, *Programs for Automatic Differentiation for the Machine BESM*, Technical Report, Institute for Precise Mechanics and Computation Techniques, Academy of Science, Moscow, 1959.

[7] C. Bennett, *Logical reversibility of computation*, IBM Journal of Research and Development, Vol. 17, 1973, 525–532.

[8] C. H. Bischof, A. Carle, G. F. Corliss, A. Griewank, and P. D. Hovland, *ADIFOR: Generating derivative codes from Fortran programs*, Scientific Programming, Vol. 1, 1992, 11–29.

[9] C. H. Bischof, G. F. Corliss, and A. Griewank, *Structured second- and higher-order derivatives through univariate Taylor series*, Optimization Methods and Software, Vol. 2, 1993, 211–232.

[10] C. H. Bischof, L. Green, K. Haigler, and T. Knauff, *Calculation of sensitivity derivatives for aircraft design using automatic differentiation*, in Proceedings of the 5th AIAA/NASA/USAF ISSMO Symposium on Multidisciplinary Analysis and Optimization, AIAA 94-4261, American Institute of Aeronautics and Astronautics, Reston, VA, 1994, 73–84.

[11] C. H Bischof and A. Griewank, *Computational differentiation and multidisciplinary design*, in Inverse Problems and Optimization Design in Industry, H. Engl and J. McLaughlin, eds., Teubner Verlag, Stuttgart, 1994, 187–211.

[12] C. H. Bischof and M. R. Haghighat, *Hierarchical approaches to automatic differentiation*, in Computational Differentiation: Techniques, Applications and Tools, M. Berz, C. Bischof, G. Corliss, and A. Griewank, eds., SIAM, Philadelphia, PA, 1996, 83–94.

[13] C. H. Bischof, P. Khademi, A. Mauer, and A. Carle, *ADIFOR* 2.0: *Automatic differentiation of Fortran 77 programs*, IEEE Computational Science and Engineering, Vol. 3, 1996, 18–32.

[14] C. H. Bischof, B. Lang, and A. Vehreschild, *Automatic differentiation for MATLAB programs*, Proceedings in Applied Mathematics and Mechanics, Vol. 2, 2003, 50–53.

[15] C. H. Bischof, H. Martin Bücker, B. Lang, A. Rasch, and A. Vehreschild, *Combining source transformation and operator overloading techniques to compute derivatives for MATLAB programs*, in Proceedings of the Second IEEE International Workshop on Source Code Analysis and Manipulation (SCAM 2002), IEEE Computer Society, Washington, DC, 2002, 65–72.

[16] C. H. Bischof, L. Roh, and A. Mauer-Oats, *ADIC—an extensible automatic differentiation tool for ANSI-C*, Software: Practice and Experience, Vol. 27, 1997, 1427–1456.

[17] C. G. Broyden, *A class of methods for solving nonlinear simultaneous equations*, Mathematics of Computation, Vol. 19, 1965, 577–593.

[18] C. G. Broyden, *The convergence of an algorithm for solving sparse nonlinear systems*, Mathematics of Computation, Vol. 25, 1971, 285–294.

[19] M. Bücker and A. Rasch, *Modeling the performance of interface contraction*, ACM Transactions on Mathematical Software, Vol. 29, 2003, 440–457.

[20] D. G. Cacuci, C. F. Weber, E. M. Oblow, and J. H. Marable, *Sensitivity theory for general systems of nonlinear equations*, Nuclear Science and Engineering, Vol. 88, 1980, 88–110.

[21] L. Capriotti and M. Giles, *Adjoint Greeks made easy*, Risk, 2012, 96–102.

[22] L. Capriotti and M. Giles, *Fast Correlation Greeks by Adjoint Algorithmic Differentiation*, 2010. Available at SSRN: http://ssrn.com/abstract-1587822

[23] L. Capriotti and M. Giles, *Fast correlation Greeks by adjoint algorithmic differentiation*, Risk, 2010, 79–83.

[24] L. Capriotti and S. Lee, *Adjoint credit risk management*, Risk, to appear.

[25] L. Capriotti, S. Lee, and M. Peacock, *Real time counterparty credit risk management in Monte Carlo*, Risk, 2011.

[26] Cayuga Research, *ADMAT 2.0 User's Guide*, http://www.cayugaresearch.com/admat.html.

[27] Z. Chen and P. Glasserman, *Fast pricing of basket default swaps*, Operations Research, Vol. 56, 2008, 286–303.

[28] Z. Chen and P. Glasserman, *Sensitivity estimates for portfolio credit derivatives using Monte Carlo*, Finance and Stochastics, Vol. 12, 2008, 507–540.

[29] B. Christianson, *Cheap Newton steps for optimal control problems: Automatic differentiation and Pantoja's algorithm*, Optimization Methods and Software, Vol. 10, 1999, 729–743.

[30] B. Christianson, *A self-stabilizing Pantoja-like indirect algorithm for optimal control*, Optimization Methods and Software, Vol. 16, 2001, 131–149.

[31] T. F. Coleman and J. Y. Cai, *The cyclic coloring problem and estimation of sparse Hessian matrices*, SIAM Journal on Algebraic and Discrete Methods, Vol. 7, 1986, 221–235.

[32] T. F. Coleman and G. F. Jonsson, *The efficient computation of structured gradients using automatic differentiation*, SIAM Journal on Scientific Computing, Vol. 20, 1999, 1430–1437.

[33] T. F. Coleman, Y. Li, and A. Verma, *Reconstructing the unknown local volatility function*, Journal of Computational Finance, Vol. 2, 1999, 77–102.

[34] T. F. Coleman, Y. Li, and C. Wang, *Stable local volatility function calibration using spline kernel*, Computational Optimization and Applications, Vol. 55, 2013, 375–702. DOI 10.1007/s10589-013-9543-x.

[35] T. F. Coleman and A. Liao, *An efficient trust region method for unconstrained discrete-time optimal control problems*, Journal of Computational Optimization and Applications, Vol. 4, 1993, 47–66.

[36] T. F. Coleman and J. J. Moré, *Estimation of sparse Hessian matrices and graph coloring problems*, Math Programming, Vol. 28, 1984, 243–270.

[37] T. F. Coleman and J. J. Moré, *Estimation of sparse Jacobian matrices and graph coloring problems*, SIAM Journal of Numerical Analysis, Vol. 20, 1983, 187–209.

[38] T. F. Coleman, F. Santosa, and A. Verma, *Semi-automatic differentiation*, in Computational Methods for Optimal Design and Control, Progress in Systems and Control Theory, Vol. 24, Birkhäuser Boston, Boston, MA, 1998, 113–126.

[39] T. F. Coleman and A. Verma, *ADMIT-1: Automatic differentiation and MATLAB interface toolbox*, ACM Transactions on Mathematical Software, Vol. 26, 2000, 150–175.

[40] T. F. Coleman and A. Verma, *The efficient computation of sparse Jacobian matrices using automatic differentiation*, SIAM Journal on Scientific Computing, Vol. 19, 1998, 1210–1233.

[41] T. F. Coleman and A. Verma, *Structure and efficient Hessian calculation*, in Advances in Nonlinear Programming, Y. Yuan, ed., Springer, Berlin, 1998, 57–72.

[42] T. F. Coleman and A. Verma, *Structured and efficient Jacobian calculation*, in Computational Differentiation: Techniques, Applications, and Tools, M. Berz, C. Bischof, G. Corliss, and A. Griewank, eds., SIAM, Philadelphia, PA, 1996, 149–159.

[43] T. F. Coleman, X. Xiong, and W. Xu, *Using directed edge separators to increase efficiency in the determination of Jacobian matrices via automatic differentiation*, in Recent Advances in Algorithmic Differentiation, Lecture Notes in Computational Science and Engineering Vol. 87, Springer, Berlin, 2012, 209–219.

[44] T. F. Coleman and W. Xu, *Fast (structured) Newton computations*, SIAM Journal on Scientific Computing, Vol. 31, 2008, 1175–1191.

[45] T. F. Coleman and W. Xu, *Parallelism in structured Newton computations*, in Parallel Computing: Architectures, Algorithms and Applications, Proceedings of the International Conference (ParCo 2007), C. Bischof, M. Bücker, P. Gibbon, G. Joubert, T. Lippert, B. Mohr, F. Peters, eds., 2007, 295–302.

[46] B. Dauvergne and L. Hascoët, *The data-flow equations of checkpointing in reverse automatic differentiation*, in Proceedings of ICCS 2006, Springer, Berlin, 2006, 566–573.

[47] J. E. Dennis, Jr. and R. B. Schnabel, *Numerical Methods for Unconstrained Optimization and Nonlinear Equations*, Classics in Applied Mathematics, Vol. 16, SIAM, Philadelphia, PA, 1996.

[48] S. A. Forth, *An efficient overloaded implementation of forward mode automatic differentiation in MATLAB*, ACM Transactions on Mathematical Software, Vol. 32, 2006, 195–222.

[49] A. H. Gebremedhin, A. Tarafdar, F. Manne, and A. Pothen, *New acyclic and star coloring algorithms with applications to computing Hessians*, SIAM Journal on Scientific Computing, Vol. 29, 2007, 1042–1072.

[50] A. H. Gebremedhin, A. Pothen, A. Tarafdar, and A. Walther, *Efficient computation of sparse Hessians using coloring and automatic differentiation*, INFORMS Journal on Computing, Vol. 21, 2009, 209–223.

[51] A. H. Gebremedhin, F. Manne, and A. Pothen, *What color is your Jacobian? Graph coloring for computing derivatives*, SIAM Review, Vol. 47, 2005, 629–705.

[52] M. Giles and P. Glasserman, *Smoking adjoints: Fast Monte Carlo Greeks*, Risk, Vol. 19, 2006, 88–92.

[53] P. E. Gill, W. Murray, and M. H. Wright, *Practical Optimization*, Academic Press, New York, 1982.

[54] R. M. Gower and M. P. Mello, *A new framework for the computation of Hessians*, Optimization Methods and Software, Vol. 27, 2012, 251–273.

[55] A. Griewank, *Some bounds on the complexity gradients*, in Complexity in Nonlinear Optimization, P. Pardalos, ed., World Scientific Publishing, River Edge, NJ, 1993, 128–161.

[56] A. Griewank, *Who invented the reverse mode of differentiation?*, in Optimization Stories, 21st International Symposium on Mathematical Programming, Martin Grotschel, ed., SPARC, 2012, 389–400.

[57] A. Griewank and G. F. Corliss, eds., *Automatic Differentiation of Algorithms: Theory, Implementation and Applications*, SIAM, Philadelphia, PA, 1991.

[58] A. Griewank, D. Juedes, and J. Utke, *Algorithm 755: ADOL-C: a package for the automatic differentiation of algorithms written in C/C++*, ACM Transactions on Mathematical Software, Vol. 22, 1996, 131–167.

[59] A. Griewank and A. Walther, *Algorithm 799: Revolve: An implementation of checkpointing for the reverse or adjoint mode of computational differentiation*, ACM Transactions on Mathematical Software,Vol. 2, 2000, 19–45.

[60] A. Griewank and A. Walther, *Evaluating Derivatives: Principles, and Techniques of Algorithmic Differentiation*, 2nd ed., SIAM, Philadelphia, PA, 2008.

[61] C. Homescu, *Adjoints and Automatic (Algorithmic) Differentiation in Computational Finance*, 2011. Available at SSRN: http://ssrn.com.abstract-1828503.

[62] W. W. Hager and H. Zhang, *A survey of nonlinear conjugate gradient methods*, Pacific Journal of Optimization, Vol. 2, 2006, 35–58.

[63] P. Hovland, C. Bischof, D. Spiegelman, and M. Casella, *Efficient derivative codes through automatic differentiation and interface contraction: An application in biostatistics*, SIAM Journal on Scientific Computing, Vol. 18, 1997, 1056–1066.

[64] M. Hasan, *DSJM: A Software Toolkit for Direct Determination of Sparse Jacobian Matrices*, Master's thesis, University of Lethbridge, Alberta, Canada, 2011.

[65] L. Hascoet and M. Arayapolo, *Enabling user-driven checkpointing strategies in reverse-mode automatic differentiation*, in Proceedings of the European Conference on Computational Fluid Dynamics, P. Wesseling, E. Oñate, and J. Périaux, eds., 2006, 1–19.

[66] M. Iri, *History of automatic differentiation and rounding estimation*, in Automatic Differentiation of Algorithms: Theory, Implementation and Application, A. Griewank and G. F. Corliss, eds., SIAM, Philadelphia, PA, 1991, 1–16.

[67] M. Iri, and K. Tanabe, eds., *Mathematical Programming: Recent Developments and Applications*, Kluwer Academic Publishers, Dordrecht, 1989.

[68] M. Iri, T. Tsychiya, and M. Hoshi, *Automatic computation of partial derivatives and round error estimates with applications to large-scale systems of nonlinear equations*, Journal of Computational and Applied Mathematics, Vol. 24, 1988, 365–392.

[69] M. Joshi and C. Yang, *Efficient Greek estimation in generic swap-rate market models*, Algorithmic Finance, Vol. 1, 2011, 17–33.

[70] M. Joshi and C. Yang, *Fast and accurate pricing and hedging of long-dated CMS spread options*, International Journal of Theoretical and Applied Finance, Vol. 13, 2010, 839–865.

[71] M. Joshi and C. Yang, *Fast Delta computations in the swap-rate market model*, Journal of Economic Dynamics and Control, Vol. 35, 2011, 764–775.

[72] C. Kaebe, J. H. Maruhn, and E. W. Sachs, *Adjoint based Monte Carlo calibration of financial market models*, Journal of Finance and Stochastics, Vol. 13, 2009, 351–379.

[73] R. Lagrada and S. Osher, *Reconciling differences*, Risk, Vol. 10, 1997, 79–83.

[74] M. Leclerc, Q. Liang, and I. Schneider, *Interest rates—fast Monte Carlo Bermudan Greeks*, Risk, Vol. 22, 2009, 84–87.

[75] S. Linnainmaa, *Taylor expansion of the accumulated rounding error*, BIT, Vol. 16, 1976, 146–160.

[76] MapleSoft, www.maplesoft.com.

[77] The MathWorks Inc., C/C++, Fortran, and Python API Reference, http://www.mathworks.com/access/helpdesk/help/pdf_doc/matlab/apiref.pdf.

[78] J. J. Moré, B. S. Grabow, and K. E. Hillstrom, *Testing unconstrained optimization software*, ACM Transactions on Mathematical Software, Vol. 7, 1981, 17–41.

[79] U. Naumann, *The Art of Differentiating Computer Programs: An Introduction to Algorithmic Differentiation*, Software, Environments, and Tools, Vol. 24, SIAM, Philadelphia, 2012.

[80] U. Naumann, *Reducing the memory requirement in reverse mode automatic differentiation by solving TBR flow equations*, in Computational Science—ICCS Lecture Notes in Computer Science, Vol. 2330, 2002, 1039–1048.

[81] Open AD, www.mcs.anl.gov/OpenAD.

[82] J. M. Ortega and W. C. Rheinboldt, *Iterative Solution of Nonlinear Equations in Several Variables*, Classics in Applied Mathematics, Vol. 30, SIAM, Philadelphia, PA, 2000.

[83] J. F. A. de O. Pantoja, *Differential dynamic programming and Newton's method*, International Journal of Control, Vol. 47, 1988, 1539–1553.

[84] L. B. Rall, *Computational Solution of Nonlinear Operator Equations*, John Wiley, New York, 1969.

[85] L. B. Rall, *Automatic Differentiation: Techniques and Applications*, Lecture Notes in Computer Science, Vol. 120, Springer-Verlag, Berlin, 1981.

[86] L. B. Rall, *Differentiation in Pascal-SC: Type GRADIENT*, ACM Transactions on Mathematical Software, Vol. 10, 1984, 161–184.

[87] L. B. Rall, *The arithmetic of differentiation*, Mathematics Magazine, Vol. 59, 1986, 275.

[88] L. B. Rall, *Differentiation arithmetics*, in Computer Arithmetic and Self-Validating Numerical Methods, C. Ullrich, ed., Academic Press, New York, 1990, 73–90.

[89] L. B. Rall, *Gradient computation by matrix multiplication*, in Applied Mathematics and Parallel Computing, H. C. Fischer, B. Riedmueller, and S. Schaefller, eds., Physica-Verlag, Berlin, 1996, 233–240.

[90] L. B. Rall and G. F. Corliss, *An introduction to automatic differentiation*, in Computational Differentiation: Technique, Applications, and Tools, M. Berz, C. H., Bischof, G. F. Corliss, and A. Griewank, eds., SIAM, Philadelphia, PA, 1996, 1–18.

[91] G. M. Ostrovskii, Y. M. Volin, and W. W. Borisov, *Über die Berechnung von Ableitungen*, Wissenschaftliche Zeitschrift der Technischen Hochschule für Chemie, Vol. 13, 1971, 382–384.

[92] B. Speelpenning, *Compiling Fast Partial Derivatives of Functions Given by Algorithms*, Ph.D. thesis, Department of Computer Science, University of Illinois at Urbana–Champaign, 1980.

[93] S. Stamatiadis, R. Prosmiti, and S. C. Farantos, *auto_deriv: Tool for automatic differentiation of a Fortran code*, Computer Physics Communications, Vol. 127, 2000, 343–355.

[94] C. W. Straka, *ADF 95: Tool for automatic differentiation of a FORTRAN code designed for large numbers of independent variables*, Computer Physics Communications, Vol. 168, 2005, 123–139.

[95] A. Verma, *ADMAT: Automatic differentiation MATLAB using object-oriented methods*, in Proceedings of the SIAM Workshop on Object Oriented Methods for Interoperable Scientific and Engineering Computing, SIAM, Philadelphia, PA, 1998, 174–183.

[96] Y. M. Volin and G. M. Ostrovskii, *Automatic computation of derivatives with the use of the multilevel differentiating technique: Algorithmic basis*, Computers and Mathematics with Applications, Vol. 11, Bora Ųcer 1985, 1099–1114.

[97] A. Walther, *Computing sparse Hessians with automatic differentiation*, ACM Transactions on Mathematics and Software, Vol. 34, 2008, 1–15.

[98] M. Wang, A. H. Gebremedhin, and A. Pothen, *An efficient automatic differentiation algorithm for Hessians: Working with live variables*, in Proceedings of the Sixth SIAM Workshop on Combinatorial Scientific Computing (CSC14), Lyon, France, 2014.

[99] G. Wanner, *Integration gewöhnlicher Differentialgleichungen*, Lie-Reihen (mit Programmen), Runge-Kutta-Methoden, B.I-Hochschulskripten, no. 831/831a, Bibliographisches Institut, Mannheim-Zürich, Germany, 1969.

[100] R. E. Wengert, *A simple automatic derivative evaluation program*, Communications of the ACM, Vol. 7, 1964, 463–464.

[101] P. J. Werbos, *Application of advances in nonlinear sensitivity analysis*, in System Modeling and Optimization: Proceedings of the 19th IFIP Conference, New York, R. F. Drenick and F. Kozin, eds., Lecture Notes in Control and Information Sciences, Vol. 38, Springer, New York, 1982, 762–770.

[102] W. Xu, X. Chen, and T. Coleman, *The efficient application of automatic differentiation for computing gradients in financial applications*, in press.

[103] W. Xu and T. F. Coleman, *Efficient (partial) determination of derivative matrices via automatic differentiation*, SIAM Journal on Scientific Computing, Vol. 35, 2013, A1398–A1416.

[104] W. Xu, S. Embaye, and T. F. Coleman, *Efficient Computation of Derivatives, and Newton Steps, for Minimization of Structured Functions Using Automatic Differentiation*, in press.

[105] W. Xu and W. Li, *Efficient preconditioners for Newton-GMRES method with application to power flow study*, IEEE Transaction on Power Systems, Vol. 28, 2013, 4173–4180.

Index